T0076118

DZHEMILEV REACTION IN ORGANIC AND ORGANOMETALLIC SYNTHESIS

CHEMISTRY RESEARCH AND APPLICATIONS SERIES

Applied Electrochemistry
Vijay G. Singh (Editor)
2010. ISBN: 978-1-60876-208-8

Heterocyclic Compounds: Synthesis, Properties and Applications
Kristian Nylund and Peder Johansson (Editors)
2010. ISBN: 978-1-60876-368-9

Influence of the Solvents on Some Radical Reactions
*Gennady E. Zaikov, Roman G. Makitra, Galina G. Midyana and Lyubov.I
Bazylyak (Editors)*
2010. ISBN: 978-1-60876-635-2

Dzhemilev Reaction in Organic and Organometallic Synthesis
Vladimir A.D'yakonov
2010. ISBN 978-1-60876-683-3

CHEMISTRY RESEARCH AND APPLICATIONS SERIES

DZHEMILEV REACTION IN ORGANIC AND ORGANOMETALLIC SYNTHESIS

VLADIMIR A. D'YAKONOV

Nova Science Publishers, Inc.
New York

Copyright © 2010 by Nova Science Publishers, Inc.

All rights reserved. No part of this book may be reproduced, stored in a retrieval system or transmitted in any form or by any means: electronic, electrostatic, magnetic, tape, mechanical photocopying, recording or otherwise without the written permission of the Publisher.

For permission to use material from this book please contact us:
Telephone 631-231-7269; Fax 631-231-8175
Web Site: http://www.novapublishers.com

NOTICE TO THE READER

The Publisher has taken reasonable care in the preparation of this book, but makes no expressed or implied warranty of any kind and assumes no responsibility for any errors or omissions. No liability is assumed for incidental or consequential damages in connection with or arising out of information contained in this book. The Publisher shall not be liable for any special, consequential, or exemplary damages resulting, in whole or in part, from the readers' use of, or reliance upon, this material.

Independent verification should be sought for any data, advice or recommendations contained in this book. In addition, no responsibility is assumed by the publisher for any injury and/or damage to persons or property arising from any methods, products, instructions, ideas or otherwise contained in this publication.

This publication is designed to provide accurate and authoritative information with regard to the subject matter covered herein. It is sold with the clear understanding that the Publisher is not engaged in rendering legal or any other professional services. If legal or any other expert assistance is required, the services of a competent person should be sought. FROM A DECLARATION OF PARTICIPANTS JOINTLY ADOPTED BY A COMMITTEE OF THE AMERICAN BAR ASSOCIATION AND A COMMITTEE OF PUBLISHERS.

LIBRARY OF CONGRESS CATALOGING-IN-PUBLICATION DATA

D'yakonov, Vladimir A.
  Dzhemilev reaction in organic and organometallic synthesis / Vladimir A. D'yakonov.
    p. cm.
  Includes bibliographical references and index.
  ISBN 978-1-60876-683-3 (softcover)
  1. Organometallic compounds--Synthesis. 2. Transition metals. 3. Phase-transfer catalysis. I. Title.
  QD411.7.S94D53 2009
  547.2--dc22
                    2009044324

Published by Nova Science Publishers, Inc. ✛ New York

CONTENTS

Preface   vii

Chapter 1   Introduction   1

Chapter 2   History of Discovery   3

Chapter 3   Ethylmagnesiation of Olefins Catalyzed by Zr
              Complexes   13

Chapter 4   Catalytic Cycloalumination and Cyclomagnesiation
              of Olefins, Acetylenes, and Allenes Mediated by Mg
              or Al Alkyls   19

Chapter 5   On Mechanism of Olefin Ethylmagnesiation and
              Cycloalumination Catalyzed by Zr Complexes   59

Chapter 6   Synthesis of Macro Metallacarbocycles through the
              Catalytic Cyclometalation Reaction of Unsaturated
              Compounds   69

Conclusion   75

Bibliography   77

Index   93

PREFACE

The present manuscript surveys the literary data published over the last 15–20 years on synthesis and study of properties of cyclic and acyclic Mg- and Al-organic compounds using the catalytic olefin ethylmagnesiation reaction as well as the cycloalumination and cyclomagnesiation reactions of olefin, acetylene, and allene (known in world literature as Dzhemilev reaction) to obtain new classes of cyclic and acyclic organometallic compounds (OMCs) namely aluminacyclopropanes, aluminacyclopropenes, aluminacyclopentanes, aluminacyclopentenes, aluminacyclopenta-2,4-dienes, magnesacyclopentanes, magnesacyclopentenes, magnesacyclopenta-2,4-dienes, and also (1,2 or 1,4)-bis(Mg or Al)ethylenes and butyl-2-enes.

The history of the catalytic ethylmagnsiation and cyclometalation reactions as well as discovered by Professor Dzhemilev phenomenon of catalytic replacement atoms of transition metals (Ti, Zr, Co) in metallacarbocycles by nontransition metal atoms (Mg, Al, Zn) to produce the above OMCs is also described.

Modern understanding of the mechanism of these reactions including structure of the key intermediates responsible for the formation of the target OMC is considered.

The main manuscript content covers information on application of the above reactions for the synthesis of various cyclic and acyclic OMCs based on Mg, Zn and Al. The unique capabilities of one pot techniques due to Dzhemilev reaction to produce carbo- and heterocyclic compounds employing unsaturated and simplest organomagnesium and organoaluminum compounds are demonstrated. The non-trivial approaches to the synthesis of various spiranes and gigantic metallacarbo- and heterocycles as well as the prospects for the further development of research in this area are also represented in this work.

Chapter 1

INTRODUCTION

The first five-membered metallacarbocycles based on transition metals (Fe, Co, Rh) have been discovered in the fifties of the last century [1–4]. From that moment the investigators of the research centers in different countries have begun intensive studying of the properties of these unique compounds [5–7]. Further researches in this direction have led to wide application of transition-metal based metallacycles in the synthesis of five-membered, six-membered, and macroheterocycles with N, O, S, Se, Si, P, Ge, and Sn atoms in the ring and other useful substances.

In 1970, with a synthesis of zirconacyclopentadiene, chemistry of transition-metal based metallacycles has received its second birth. In literature, from that moment, there is a kind of boom on the synthesis of metallacycles based on group IV–X transition metal and their transformations to carbo- and heterocyclic compounds [8].

Meanwhile in the synthesis of nontransition-metal based metallacarbocycles such as alumina- and magnesacyclopentanes only scattered data were available [9–17]. Methods for their preparations were based mainly on the thermal intramolecular hydroalumination reactions of 1,3-dienes or on the interaction between the latters and high active Mg* (Rieke magnesium) and the shift of the Schlenk equilibrium from acyclic to cyclic form of 1,4-dimagnesium reagent.

Despite the paucity of representatives of this class of compounds, their great synthetic potential was obvious. High reactivity of metal–carbon bond in nontransition metal based metallacarbocycles in comparison with those containing transition metals combined with a relatively high stability of these compounds under normal conditions together with the possibility of conducting synthetic transformations in one preparative stage allow to consider the investigations in the field of a synthesis of this class of metallacycles as one of the most promising directions in organic and organometallic synthesis.

Nevertheless, even after several decades after their discovery neither transition metal based metallacarbocycles nor cyclic OMC containing nontransition metals, have been widely used in synthetic practice owing to some limitations such as, in the first case, high cost and low availability of the starting reagents, while in the latter case, lack of efficient preparative methods for a synthesis.

Taking into consideration all of the above, one can say without exaggeration that the most important discoveries of the last 15−20 years in chemistry of organometallic compounds include the discovery of the catalytic cyclometalation reaction of unsaturated compounds using available Mg and Al alkyls and halogen alkyls in the presence of complex catalysts based on Zr, Ti, Hf and Co to obtain three-membered, five-membered and macrocyclic organometallic compounds.

Discovered by Professor U.M. Dzhemilev phenomenon of catalytic replacement of transition metal atoms in metallacarbocycles by nontransition metal atoms to afford the corresponding organometallic compounds provided basis of this pioneer research. As a result, the new class of organometallic reactions has been elaborated. The latters have gained worldwide recognition and are used in synthetic practice as *Dzhemilev reactions*. Further development of these reactions in turn initiated a new direction in organoaluminum and organomagnesium synthesis, *viz.*, one pot design under mild conditions of new and different structure types of cyclic and acyclic OMCs.

This work attempts to represent in chronological sequence the major milestones and prerequisites prior to discovery of the catalytic cyclometalation reaction as well as provides examples of its use in synthesis of novel nontransition metal based metallacycles via cyclometalation of olefins, dienes and acetylene mediated by Grignard reagents and haloalkylalanes in the presence of Zr, Ti and Co complex catalysts. It also demonstrates the unique features of the reaction in "one-pot" transformations of unsaturated compounds into carbo- and heterocycles through generated *in situ* cyclic OMCs.

The author also considers the strategy and prospects for further *Dzhemilev reaction* progress.

Chapter 2

HISTORY OF DISCOVERY

History of discovery of the catalytic cyclometalation reaction of unsaturated compounds, which allow synthesizing five-membered OMCs from α-olefins, is very interesting and instructive. The discovery of the above reaction was preceded by the studies carried out in U. M. Dzhemilev's scientific group on linear dimerization and codimerizattion of 1,3-dienes and α-olefins in the presence of Zr catalysts [18−21]. During dimerization reaction of α-olefins affected by Zr(OBu)₄–Et₂AlCl catalyst one could observe together with the target methylene alkanes **1** in every experience the formation of compound **2** as the minor β-ethylation product (~ 5%) from initial α-olefins (Scheme 1).

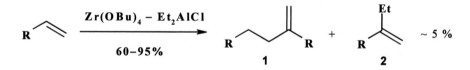

Scheme 1.

The increase in concentration of Et₂AlCl in the catalyst composition caused the increase in the yield of the β-ethylation product **2** up to ~ 90% while using stoichiometric amounts of diethylaluminum chloride [22, 23] (Scheme 2).

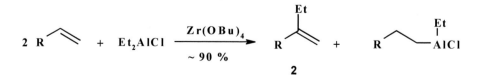

Scheme 2.

While studying the influence of the nature and structure of ligands surrounding the central atom of a catalyst as well as the search of metal complex catalysts based on other transition metals able to promote the dimerization reaction of α-olefins, it was found [24] that the replacement of Zr(OBu)$_4$ or (BuO)$_n$ZrCl$_{4-n}$ by Ti(OBu)$_4$, TiCl$_4$ or Cp$_2$TiCl$_2$ complex catalysts caused reductive 1,2-carboalumination of initial α-olefins including those with the use of AlEt$_3$ (Scheme 3).

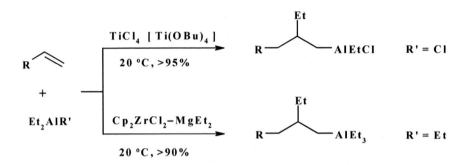

Scheme 3.

It was suggested that the above reaction proceeds through the generation of zircona- or titanacarbocycles as the key intermediates according to the following scheme:

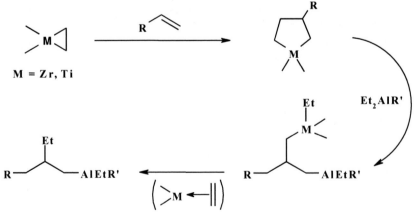

Scheme 4.

By analogy with the carboalumination reaction, in 1983, for the first time, there was published regioselective 1,2-**carbomagnesiation** of α-olefins with nonactivated double bond (*Dzhemilev reaction* [25–28]), including those with substituents containing various functional groups [29]. According to the work [30], zirconacyclopropanes and zirconacyclopentanes were considered as the key intermediates, whose sequential transformations under reaction conditions led to the target 1,2-carbomagnesiation products [31] (Scheme 5).

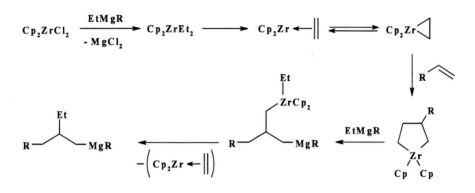

Scheme 5.

From the examples given above, one can conclude that the catalytic 1,2-carbometallation of olefins with Mg and Al alkyl derivatives affected by Zr and Ti complexes occurs *via* the generation of intermediate transition metal metallacarbocycles, which are responsible for the formation of target organometallic compounds, under reaction conditions.

The formation and subsequent transformation of zircona- and titanacarbocycles under conditions of 1,2-carbomagnesiation or 1,2-carboalumination were accomplished due to the unique ability of Zr and Ti alkyl or cycloalkyl complexes to transfer their π- and σ-bound ligands onto nontransition metal atoms and *vice versa* [32–34].

At the beginning of the 80s, based on the above results, Professor Dzhemilev was the first to announce an idea about the possibility of synthesising magnesa- and aluminacycloalkanes by the catalytic cyclometallation of olefins with Mg or Al alkyl derivatives in the presence of Zr or Ti complexes.

The essence of the advanced idea consisted in the ability of the coordinatively unsaturated Zr or Ti complexes to coordinate olefins to give donor–acceptor complexes in accordance with the *Dewar–Chatt–Dunkanson* model [35] leading thus to the activation of initial olefins with the subsequent intramolecular oxidative cyclization of these latter to form three- and five-membered transition metal metallacarbocycles (Scheme 6):

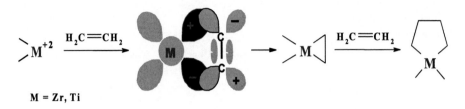

M = Zr, Ti

Scheme 6.

Via the interaction between the obtained transition metal metallacycles and nontransition metal alkyls or alkyl halides, for example, AlR_3 or $MgRR'$, one could hope to replace transition metal atoms in metallacarbocycles by the atoms of nontransition metals to obtain the corresponding nontransition metal metallacycles according to the following Scheme:

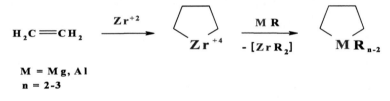

M = Mg, Al
n = 2-3

Scheme 7.

Certainly, it remained not clear if these reactions will proceed in stoichiometric or catalytic versions (Scheme 8).

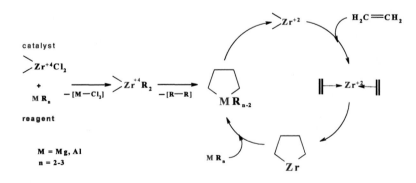

Scheme 8.

This catalytic reaction offered completely new potentials to construct in one preparative stage the reactive metallacarbocycles from olefins, as well as in prospect from acetylenes, obtaining both transition and nontransition metal metallacarbocycles.

Research in this direction has been crowned with success. In the middle of the 80s the said reactions to produce nontransition metal based metallacarbocycles (Al and Mg) were performed first in stoichiometric and then catalytic versions. In this case, Cp_2ZrCl_2 and Cp_2TiCl_2 proved to be the most active and selective catalysts.

According to the scheme below, the rapid ligand exchange occurs *via* the interaction between Cp_2ZrCl_2 and EtMgR (R = Et, Cl) or $AlEt_3$ to give Cp_2ZrEt_2, which transforms into zirconocene Cp_2Zr^{2+} as a result of β-elimination of hydrogen atoms. The latter, under reaction conditions, coordinates the molecule of ethylene or a higher olefin to form corresponding zirconacyclopropanes. The subsequent introduction of the initial olefin at the Zr–C bond leads to appropriate 3-substituted or *trans*-3,4-disubstituted zirconacyclopentanes, which in the presence of the excess of magnesium or aluminum alkyl derivatives transmetalate to give the target magnesa- or aluminacyclopentanes.

This reaction is characterised by the extremely high region and stereoselectivity allowing in one **preparative stage from** α-olefins and magnesium or aluminum alkyl derivatives to synthesize five-membered nontransition metal based metallacarbocycles (Mg, Al) in practically quantitative yields.

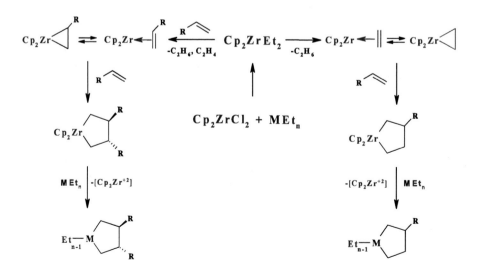

Scheme 9.

The above transformations have been initially performed in a stoichiometric version and then with the use of catalytic amounts of Zr or Ti complexes and Hf, Ta and Co as well.

The study of application boundaries of this reaction with the participation of cyclic and acyclic olefins [36–38] resulted in the development of a general catalytic method for the construction of five-membered nontransition metal based metallacarbocycles of different structures.

The first announcement concerned the possibility of the preparative synthesis of aluminacyclopentanes under catalytic reaction conditions from α-olefins and Et₃Al affected by the Cp₂ZrCl₂ catalyst has already appeared in scientific periodicals in 1989 [39], though, as the authors of the work [24] notice, these results were gained as early as 1985.

Novelty and singularity of the data obtained, *viz.*, the construction of five-membered OACs in one preparative stage from starting acyclic reagents with high selectivity (more than 95%) and quantitative yields (more than 90%), urged rigorous with a large share of self-criticism study of the given reaction for reliable structure design of the OACs obtained. As a result, 3-substituted aluminacyclopentanes (**3**) have been obtained in high yields via the interaction between α-olefins and Et₃Al in the presence of 5 mole % Cp₂ZrCl₂. The structure of cyclic OACs (**3**) was established by spectral methods as well as through chemical transformations to butane-1,4-diols (**4**) and 1,4-dideuterobutanes (**5**) as follows [39]:

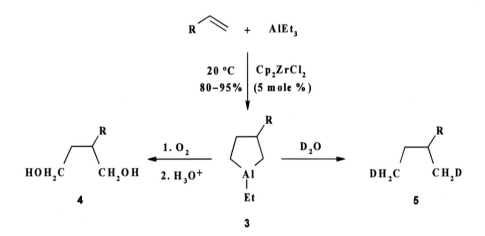

Scheme 10.

Thereby, during catalytic dimerization of α-olefins the product (**2**) identified as the minor product (~5%) further became the starting point for elaboration of the new catalytic ethylmagnesiation and cycloalumination reactions of unsaturated compounds. The developed reaction, which allows synthesizing alumina- and magnesacyclopentanes from unsaturated compounds with the aid of trialkyl- and alkylhalogenalanes, was entitled as *"the catalytic cycloalumination reaction"* [34, 37, 38, 40] and, nowadays, widely used in synthetic practice as *Dzhemilev reaction* [25].

It remained unclear whether the catalytic cyclometalation reaction is typical for olefins or it can be extended to other classes of unsaturated compounds, for example, allenes and acetylenes. In this case, one could expect to obtain transition and nontransition metal based metallacarbocycles such as zirconacyclopropenes or titanacyclopropenes and, consequently, not previously described alkylidenealuminacyclopropanes and aluminacyclopentanes, aluminacyclopropenes, aluminacyclopentenes, aluminacyclopentadienes as well as other similar nontransition metal metallacarbocycles and its acyclic analogues in the case of the thermodynamic stability.

As a result of practical realization of general schemes for the synthesis of cyclic and acyclic aluminum organic compounds (OACs), preparative methods have been developed by Dzhemilev and coworkers to synthesize aluminacyclopropanes [41,42], aluminacyclopropenes, aluminacyclopentanes, aluminacyclopentenes, aluminacyclopentadienes and 1,2-dialuminioethylenes and to study their physicochemical properties [43–45] (Scheme 11).

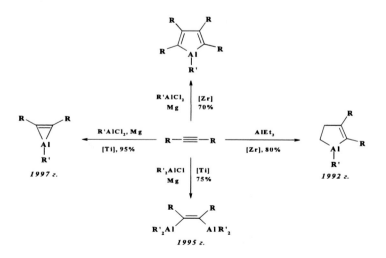

Scheme 11.

As it turned out, the above classes of OACs were stable under inert conditions and all reactions described for acyclic aluminum organic compounds were characteristic of them in most cases.

The analysis of the data obtained on the study of the catalytic cyclometallation reaction of olefins and acetylenes with the aid of Al or Mg alkyl derivatives in the presence of Zr or Ti complexes showed that all these reactions have general nature and the formation of metallacarbocycles on the basis of Al and Mg occurs in all experiments through the formation of transition metal metallacarbocycles, in particular, zirconium and titanium.

Properly, in all experiments the catalytic replacement of transition metal atoms (Zr, Ti) in the appropriate transition metal metallacarbocycles by the atoms of nontransition metals (Al, Mg) occurs to form the nontransition metal metallacarbocycles.

This reaction is characterized by high regio- and stereoselectivity, makes it possible to conduct the formation of metallacycles at room temperature and in practically quantitative yields.

The catalytic cyclometallation reaction of olefins can be represented by the following general Scheme:

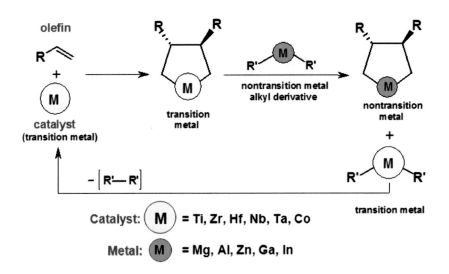

Scheme 12.

The wide and active study of the catalytic cyclometallation reaction by domestic and foreign researchers led to the development of new classes of cyclic and acyclic organometallic compounds, and to the discovery of the family of organic and organometallic reactions, which make it possible to synthesize at one preparative stage small, average and macrocyclic compounds, bifunctional monomers with substituents of assigned configuration, heterocycles and other useful synthons from simplest olefins, acetylenes and organometallic reagents (Scheme 13).

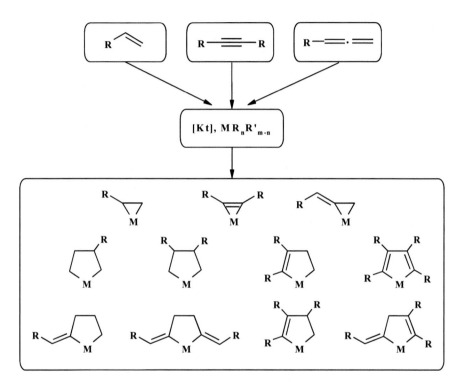

M = Mg, Al, Zn; n, m = 0-3; [Kt] = Zr, Ti, Hf, Co; R, R' = Cl, Et

Scheme 13.

Further, aspects of the catalytic ethylmagnesiation reaction of olefins and the catalytic cyclomagnesiation reactions of olefins, acetylenes and allenes will be discussed in detail.

Chapter 3

ETHYLMAGNESIATION OF OLEFINS CATALYZED BY ZR COMPLEXES

The term «carbometalation», suggested in 1978 by E. Negishi and coworkers [46] for the reaction of 1,2-addition of organometallic reagents to unsaturated compounds, later on became commonly adopted [47−50].

Only alkenes with activated double bonds can be involved in the reaction of thermal carbomagnesiation with the aid of allylmagnesium halides [47,51,52] (Scheme 14).

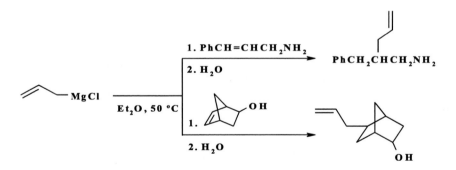

Scheme 14.

Catalysts based on titanium and other transition metal chlorides ($NiCl_2$ or $CoCl_2$) enabled catalytic carbomagnesiation of α-alkenes devoid of activated double bonds with Grignard reagents [53−55]. However, these catalysts have not found wide application because of low regio- and stereoselectivities.

Only due to discovery in 1983 in the group of Professor Dzhemilev of the catalytic ethylmagnesiation reaction of α-olefins with nonactivated C−C double bonds regioselective **carbomagnesiation** of α-alkenes has been carried out for the first time. High yields of products were obtained in the presence of a catalytic amount of Cp₂ZrCl₂ using Et₂Mg or EtMgBr. This reaction, referred to as "catalytic ethylmagnesiation", is applicable to a wide range of olefins [29, 56, 57].

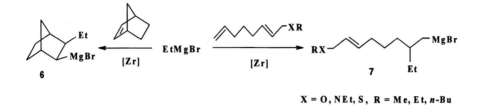

X = O, NEt, S, R = Me, Et, *n*-Bu

Scheme 15.

In particular, bicyclo[2.2.1]hept-2-ene reacts with an equimolar amount of Et₂Mg in the presence of a catalytic amount of Cp₂ZrCl₂ (20 °C, 4 h) to form compound (6), further hydrolysis of which led to 2-*exo*-ethylbicyclo[2.2.1]heptane (60–65%) (Scheme 15). The interaction between Et₂Mg and *endo*-tricyclo[5.2.1.0²,⁶]deca-3,8-diene was established to afford a mixture of regioisomeric OMCs (8) differed from each other in the position of the ethyl group relative to the double bond. Under these conditions, the organomagnesium compound (9) was obtained from bicyclo[2.2.1]hepta-2,5-diene and Et₂Mg in the ~70% yield. Hydrolysis of the latter resulted in 3-ethylnortricyclene [56] (Scheme 16).

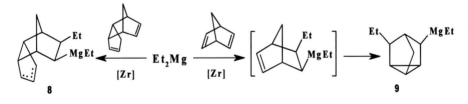

Scheme 16.

Olefins, which contain hetero atoms such as N, O or S, have been also involved in the ethylmagnesiation reaction [57–83]. Ethylmagnesiation of 2,7-octadienyl ethers and 2,7-octadienyl amines using EtMgR (R = Br, Et) and Cp₂ZrCl₂ catalyst occurs at the terminal double bond to form appropriate N- and O-containing OMC (**7**) with more than 90% yield (Scheme 15).

Ethylmagnesiation of allylic and homoallylic alcohols as well as ethers with EtMgCl (excess) in the presence of Cp_2ZrCl_2 has been studied by Hoveyda and coworkers. This reaction was highly stereoselective and afforded β-ethyl-substituted organomagnesium compounds in high yields. The structure of these OMCs was established by the analysis of its oxidation products (10) and (11) [58−61] (Scheme 17). High stereoselectivity of the ethylmagnesiation reaction, according to the authors, is due to the coordination of the heteroatom of the starting unsaturated compound to a central metal atom of the catalyst.

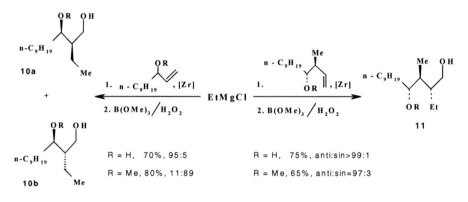

Scheme 17.

Under the effect of chiral Zr-containing catalysts asymmetric **ethylmagnesiation of** α-olefins, acyclic and cyclic allylic alcohols, ethers and amines [62–79] proceeds with high diastereo-and enantioselectivity (Scheme 18). Enantioselective formation **of C−C bonds** under the action of the chiral Zr and Ti complexes is discussed in a variety of reviews [72–75].

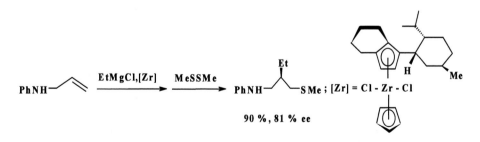

Scheme 18. Continued on next page.

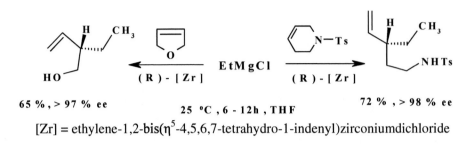

[Zr] = ethylene-1,2-bis(η^5-4,5,6,7-tetrahydro-1-indenyl)zirconiumdichloride

Scheme 18.

The enantioselective alkylation of cyclic allylic ethers or acyclic homoallylic alcohols (or ethers) with *n*-PrMgCl or *n*-BuMgCl in the presence of a catalytic amount of Cp_2ZrCl_2 **affords γ-substituted α-olefins** (Scheme 19). The regioselectivity of this reaction depends upon the structure of the starting reagents and temperature [83,84]. Typically, ether is used as a solvent, while tetrahydrofuran provides lower selectivity and yields of the products.

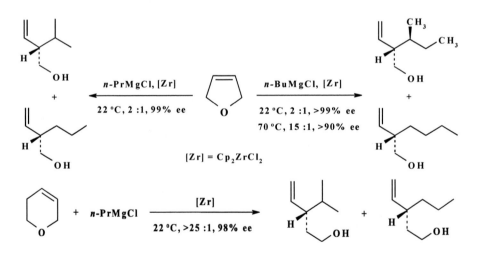

Scheme 19.

The authors of work [85] have developed the ethylmagnesiation "titanium analogue", which was used in the synthesis of substituted azabicyclo[n.1.0]alkanes. At one stage during the procedure N-benzylpyrroline was involved in the carbomagnesiation reaction using Grignard reagents in the presence of Ti complexes giving appropriate N-benzyl-N-(2-alkylbut-3-enyl) amines in high yields (Scheme 20).

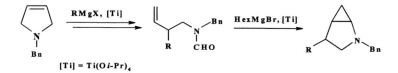

[Ti] = Ti(O*i*-Pr)₄

Scheme 20.

The olefin ethylmagnesiation reaction catalyzed by Zr complexes has also found its application in the synthesis of biologically active substances including those of natural structure [71, 80–82] (Scheme 21).

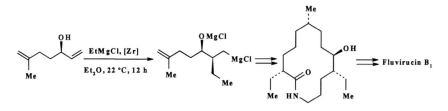

Scheme 21.

By analogy with ethylmagnesiation [29], Negishi and coworkers has succeeded in ethylzincation [123] of α-olefins using Et₂Zn and catalytic amounts of Et₂ZrCp₂ generated from Cp₂ZrCl₂ and EtMgBr. Organozinc compound (**12**) has been obtained by the method described (Scheme 22).

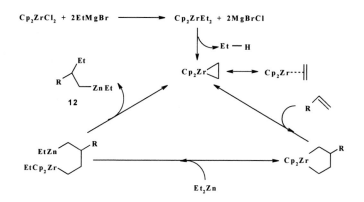

Scheme 22.

Chapter 4

CATALYTIC CYCLOALUMINATION AND CYCLOMAGNESIATION OF OLEFINS, ACETYLENES, AND ALLENES MEDIATED BY MG OR AL ALKYLS

Synthetic potential of any new reaction is summarized from a multitude of factors, the key among which are the possibility to extend this reaction to a wide range of monomers, versatility and also preparative availability of the starting reagents. As far as organometallic compounds containing high active metal−carbon bonds are concerned, one should add such factors as a wide spectrum and diversity of functional compounds, which can be obtained eventually.

This Chapter will provide insight into evolutionary development of *Dzhemilev reaction*. Its application to design and synthesis of new Al and Mg based metallacarbocycles as well as further transformations of the latters *in situ* to the new or hard-to-reach carbo- and heterocyclic, bifunctional and also known natural biologically active compounds will be demonstrated and discussed.

4.1. CATALYTIC CYCLOALUMINATION OF OLEFINS AND 1,2-DIENES TO ALUMINACYCLOPENTANES

The first announcement concerned the possibility of the preparative synthesis of the hard-to-reach five-membered aluminaccarbocycles (13) from appropriate α-olefins and Et$_3$Al in the presence of Cp$_2$ZrCl$_2$ catalyst in practically quantitative yields has appeared in 1989 [39] (Scheme 23).

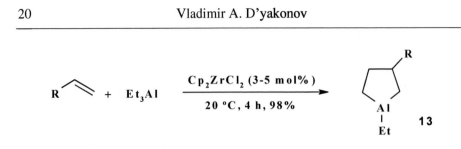

Scheme 23.

The subsequent efforts of the researchers, who first elaborated one pot procedure to convert acyclic OACs into aluminacyclopentanes, have been directed towards determination of application boundaries of the said reaction and also to the search of the catalysts able to convert α-olefins in the presence of trialkyl- and alkylhalogenalanes into corresponding substituted aluminacyclopentanes. From a number of the tested catalysts based on salts or complexes of transition metals (Cu, Mn, Cr, Ti, Zr, Hf, H, W, Mo, Fe, Cu, Ni, Pd, Rh) and widely used in metal complex catalysis only cyclopentadienyl Ti, Zr and Co complexes were testified as the most active ones to convert olefins to aluminacyclopentanes with the aid of trialkyl or alkyl halogenides with high yields and selectivity [34, 37, 38, 40, 87, 88].

Cycloalumination of α-olefins with Et_3Al in the presence of chiral Zr-containing catalysts [89,90] or cocatalysts, viz., amides or aluminum alkoxides [91] was shown to afford optically active 3-substituted aluminacyclopentanes (13), oxidation of which gave rise to optically active diols (14) (Scheme 24).

It has turned out that the reaction was sensitive to the nature of central metal atom of the catalyst. Thus, the use of Cp_2TiCl_2 as a catalyst instead of Cp_2ZrCl_2 was found to facilitate the formation of the hydroalumination products (15) [92], while the reaction of α-olefins with Et_3Al in the presence of t-BuBr and Cp_2TiCl_2 catalyst resulted in the hydroalkylation product (16) in 85–92 % yield [93] (Scheme 24).

In addition, the catalytic cycloalumination reaction is sensitive not only to the type of a catalyst but also to the nature of the solvent. Thus, the reaction of α-olefins with $AlEt_3$ in CH_3CHCl_2 solution affected by cyclopentadienyl-amidotitanium dichloride η^5-$(C_5Me_4)SiMe_2N(t$-Bu$)TiCl_2$ complex did not produce the hydroalumination (15) or hydroalkylation (16) products but was supposed to proceed through the formation of corresponding aluminacyclopentanes identified as the deuterolysis (17) and oxidation (18) products [94] (Scheme 25).

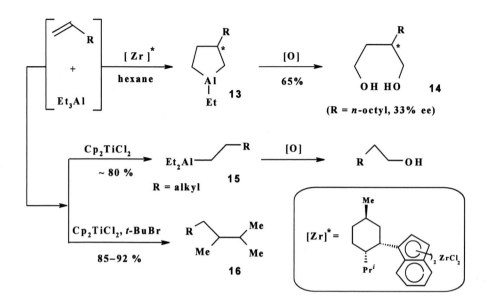

Scheme 24.

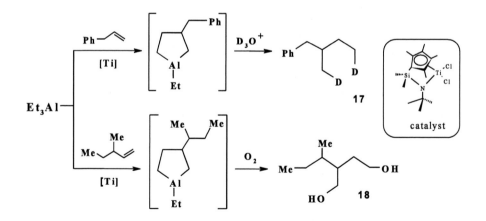

Scheme 25.

The function substituted N-, O-, and S-containing α-olefins in the presence of catalytic amounts of Cp₂ZrCl₂ enter the reaction with Et₃Al giving aluminacyclopentanes (19), in which lone electron pairs of the heteroatom such as O, N or S form donor-acceptor complexes [32]. One should not exclude the participation of disubstituted double bond in the formation of coordination environment of the aluminum atom (Scheme 26).

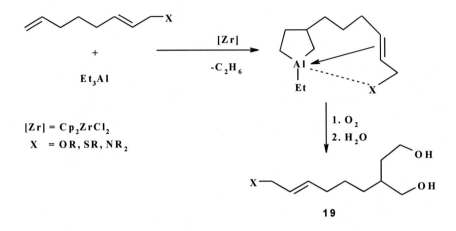

Scheme 26.

Unlike aliphatic α-olefins, which under the effect of Cp_2ZrCl_2 catalyst enter the reaction with Et_3Al yielding 1-ethyl-3-alkylaluminacyclopentanes, 1-arylolefins such as sterene, *orthor* or *para*-methylsterene under chosen conditions gave a mixture of substituted tri- (24) and five-membered (20-23) OACs at a ratio of (20):(21):(22):(23):(24) = 50:25:15:3:7 [95] (Scheme 27). The authors of the work [95] suggested that formation of cyclic OACs (20-24) occured through Zr- and Al-containing bimetallic intermediates generated from Cp_2ZrCl_2 and Et_3Al [96–99].

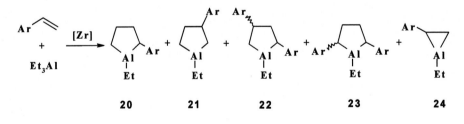

Scheme 27.

Similar results have been obtained in the course of cycloalumination of vinyl and allyl silanes with Et_3Al in the presence of Cp_2ZrCl_2 catalyst. Thus, the interaction between triethyl(vinyl)silane and Et_3Al (excess) in the presence of Cp_2ZrCl_2 as a catalyst (5 mole %, 10 h, 20 °C) was shown to afford the OAC mixture of 1-ethyl-3-(triethylsilyl)aluminacyclopentane (**25**) and 1-ethyl-2-(triethylsilyl)aluminacyclopropane (**26**) at a ratio of 6:1 (total yield ~70 %) [100] (Scheme 28). Allyl silanes unlike trialkyl(vinyl)silanes reacted more selectively to give predominantly 3-substituted aluminacyclopentanes.

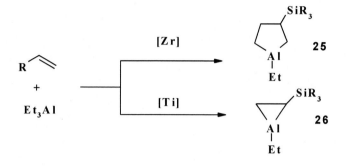

Scheme 28.

The differences observed in the cycloalumination reaction for aliphatic α-olefins, arylolefins, and vinyl silanes with the aid of Et_3Al can be explained by the nature and various structure of substituents in starting unsaturated compounds. Apparently, chemoselectivity of cyclic OAC formation is determined by the stage involving generation of Zr- and Al-containing bimetallic complexes [96–99] from zirconacarbocycles and trialkylallanes under reaction conditions.

One should believe that intermediate zirconacyclopropanes as a result of cycloalumination of aryl olefins and vinyl silanes with $AlEt_3$ in the presence of Cp_2ZrCl_2 catalyst are stabilized due to complexation of aryl and silyl substituents in starting α-olefins with central atom of the catalyst leading to the appropriate aluminacyclopropanes according to the following Scheme.

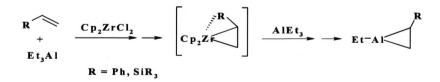

Scheme 29.

High selectivity of the olefin cycloalumination reaction has been demonstrated on the example of cycloolefins such as norbornene, norbornadiene and *exo*-dicyclopentadiene. In each experiment cycloalumination of double bond in norbornene occurred strongly stereo- selectively to afford appropriate aluminacyclopentanes (**27–29**) of *exo*-configuration [101] (Scheme 30).

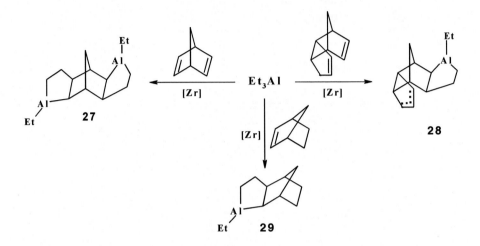

Scheme 30.

In order to develop these investigations the authors [102] have studied cycloalumination of fullerene [60] with AlEt₃ taken in excess (Scheme 31). It was shown that under chosen conditions (~23 °C, 36 h, toluene) in the presence of Cp₂ZrCl₂ catalyst the reaction occurs at a 6,6-double bond providing access to adducts with annulated to the fullerene spheroid aluminacyclopentane moieties, the number of which is dependent upon the ratio of starting reagents.

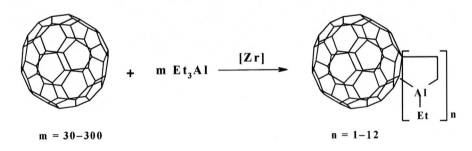

Scheme 31.

Further, the authors of the works [103] showed that together with Et₃Al in the cycloalumination reaction higher trialkylalanes R₃Al can be used. Thus, the interaction between equimolar amounts of α-olefins and higher trialkylalanes in the presence of 3 mole % of Cp_2ZrCl_2 for 12 hours at ambient temperature was found to produce selectively 1-alkyl-*trans*-3,4-dialkyl-substituted ACP (**30**) in 50–75% yield. Based on OAC (**30**) the preparative method for a synthesis of *threo*-2,3-dialkylbutane-1,4-diols (**31**) from α-olefins has been developed according the following Scheme 32.

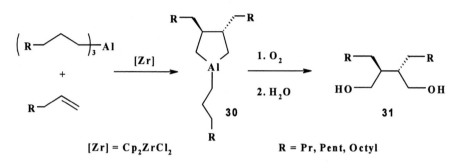

Scheme 32.

The all aforesaid allow to consider the use of pyrophoric OACs in the synthesis of aluminacyclopentanes as a principal deficiency that noticeably limits the preparative value of the said method. Our proposed method for a synthesis of 1-ethyl-*trans*-3,4-dialkylaluminacyclopentanes (**32**) by the interaction between α-olefins and $EtAlCl_2$ in the presence of metallic Mg and catalytic amounts of Cp_2ZrCl_2 at ambient temperature in tetrahydrofuran does not have this limitation [36, 104]. In the cycloalumination reaction under said conditions together with $EtAlCl_2$ one can use alkoxides, aluminum amides $RAlCl_2$ (R = OR', NR'$_2$) or $AlCl_3$ to provide (**32**) in 70–90 % yield [105] (Scheme 33). This approach was successfully applied to the preparation of substituted indacyclopentanes [106].

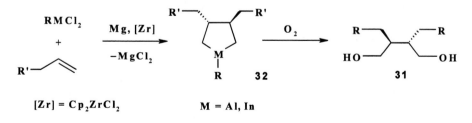

Scheme 33.

As is evident, the above reactions proceeded through generation from α-olefins and Cp_2ZrCl_2 catalyst zirconacyclopentane intermediates (**33**) [107−110], transmetalation of which with RAlCl$_2$ led to *trans*-3,4-dialkylaluminacyclopentanes (34) in high yields with high selectivity (Scheme 34).

During these investigations the authors of the works [111] extensively studied the catalytic cycloalumination reaction for various α-olefins in the presence of Cp$_2$ZrCl$_2$ catalyst and metallic Mg (THF, ~20 °C) assisted by dialkyl aluminum chlorides, alkoxides and also aluminum amides of the general formula R$_2$AlCl (R = alkyl, OR^1, NR$^1{}_2$). As a result, *threo*-2,3-dialkyl-1,4-dialuminiobutanes (35) have been synthesized in one preparative stage in 64–84% yield.

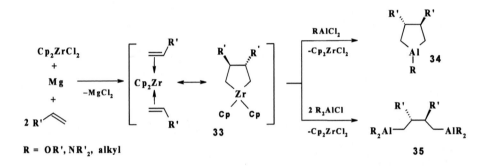

R = OR', NR'$_2$, alkyl

Scheme 34.

The direction of the given reactions was found to depend upon both the type of initial reagents and chemical nature of a catalyst. The use of Cp$_2$TiCl$_2$ catalyst instead of Cp$_2$ZrCl$_2$ promoted hydroalumination of α-olefins giving rise to ethyl dialkyl alanes (36) in 60–85% yield [112] (Scheme 35). The authors of the work [112] assumed that in this hydrogen atom transfer (HAT) reaction the solvent (THF) was appeared as hydrogen donor.

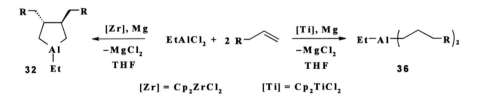

Scheme 35.

The synthesized by the said method 2,3-dialkylsubstituted 1,4-dialuminum compounds contain two asymmetrical C-2 and C-3 carbon atoms being able to form diastereomeric pair. The spectral ^{13}C NMR analysis of 1,4-dialuminio compounds (35), their the hydrolysis and deuterolysis products allowed to add these OAC to *threo* stereo isomers [111] (Scheme 36). The structure of aluminacyclopentanes as well as the structure of tri- and tetracyclic OACs, the position of substituents and its configuration have been reliably established by spectral methods [113, 114].

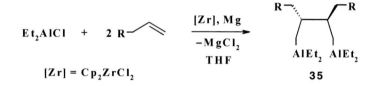

Scheme 36.

High stereoselectivity and efficiency of the preparation method to obtain *trans*-3,4-dialkylaluminacyclopentanes (32) using available fire and explosion safety reagents *viz.* RAlCl$_2$ [36, 104, 105], specify the prospect of its wide application in synthetic practice.

As it follows from the reactions given above, the methodology of catalytic cycloalumination of α-olefins with RAlCl$_2$ to *trans*-3,4-disubstituted aluminacyclopentanes does not allow to synthesize 3-substituted aluminacyclopentanes. The authors of the works [115] have succeeded in obtaining of 3-alkylsubstituted cyclic OAC (37) with RAlCl$_2$ (R = Et, OR', NR$_2$') through the combined cycloalumination of α-olefins and ethylene generated *in situ* from 1,2-dichloroethane with the aid of dihalogenalanes in the presence of metallic Mg (excess) and Cp$_2$ZrCl$_2$ catalyst in tetrahydrofuran [115] (Scheme 37). In these experiments *trans*-3,4-dialkylaluminacyclopentanes (35) were detected in minor amounts (<10 %).

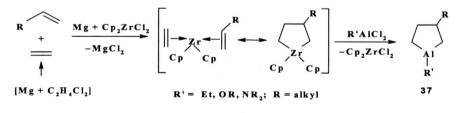

Scheme 37.

Synthesized and described above monosubstituted and *trans*-3,4-disubstituted aluminacyclopentanes are characterized by high reactivity. Thus, the Al–C bonds in aluminacyclopentanes manifested higher reactivity as compared with Al–Et bond in acyclic OACs [32]. For example, 3-alkylaluminacyclopentanes enter the reaction with α-olefins in the presence of Ti catalysts to produce 1-ethyl-3,6-dialkylaluminacycloheptanes (38) [116]. The use of allyl chlorides instead of α-olefins as well as Ni or Co catalysts instead of Zr complexes were found to promote the cleavage of aluminacyclopentane ring giving rise to acyclic allyl halogenalanes (39) [117, 118] (Scheme 38).

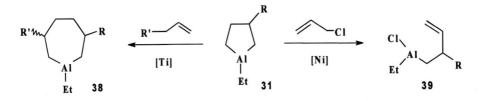

Scheme 38.

The cross-coupling reaction of substituted aluminacyclopentanes with allyl halogenides affected by CuCl catalyst is a convenient rout to the appropriate substituted hept-1-enes and *threo*-5,6-dialkylsubstituted deca-1,9-dienes [119] (Scheme 39).

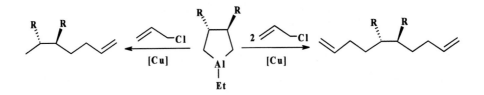

Scheme 39.

Higher reactivity of Al–C bonds in aluminacyclopentane ring as compared with Al–Et bond in acyclic OACs reveals the simple way to the synthesis of carbo- and heterocycles from α-olefins through the preliminary generation of the appropriate aluminacyclopentanes. Thus, the interaction between generated *in situ* 1-ethyl-3-alkylaluminacyclopentanes (13) and allyl chloride (excess) in the presence of Ni(acac)$_2$ catalyst in diethyl ether was shown to afford 1,1-disubstituted cyclopropanes (42) in 55–70% yield [120]. According to the Scheme below aluminacyclopentanes (13) under the effect of Ni complex catalyst undergo the intramolecular hydrogen transfer to form but-3-enyl(ethyl)aluminum hydrides (40), which further react with initial allyl halogenide giving rise to but-3-enyl(ethyl)aluminum halogenides (41). Subsequent intramolecular carboalumination resulted in the corresponding 1,1-disubstituted cyclopropanes (42) [121, 122] (Scheme 40).

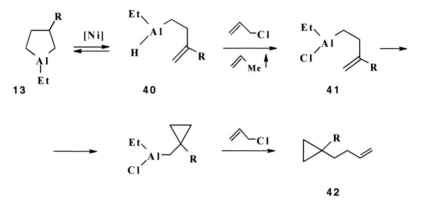

Scheme 40.

To develop these works the simple procedure to synthesize monoalkylsubstituted cyclobutanes (43) through the reaction of 3-alkylsubstituted aluminacyclopentanes (13) with allyl halogenides (1:3 molar ratio) at ambient temperature have been realized in diethyl ether in the presence of Pd(acac)$_2$–Ph$_3$P complex catalyst (5 mole %) in 60–78% yield. The reaction was accompanied by the elimination of propylene and cyclopropane (as a gaseous mixture at a ratio of 5:1) generated from allyl halogenides under the action of low valence Pd complexes [123]. Under above conditions one can obtain the corresponding *trans*-1,2-dialkylcyclobutanes with high selectivity from 1-ethyl-*trans*-3,4-dialkylaluminacyclopentanes (32) [124]. The reaction is very sensitive to the nature of both allyl compound and central atom of a catalyst. Thus, in the presence of ~3,5-fold excess of allyl acetate and Ni(acac)$_2$ catalyst (5 mole %) in THF at room temperature 1-ethyl-3-alkylaluminacyclopentanes (13) converted into 2-alkylbuta-1,3-dienes (44) in 71–76 % yield [125] (Scheme 41).

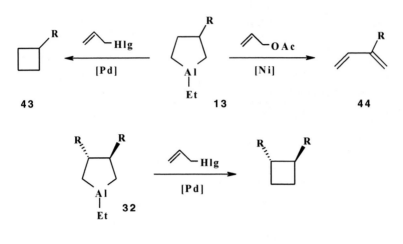

[Pd] = Pd(acac)₂-Ph₃P ;   [Ni] = Ni(acac)₂ ;   R = alkyl

Scheme 41.

The interaction between cyclooctasulfane (S_8) or selenium (Se) and 3-alkyl-substituted aluminacyclopentanes (13) at a ratio of ~3:1 in benzene (80 °C, 6 h) were shown to afford 3-alkyltetrahydrothiophenes (45) or tetrahydroselenophenes (46) [126]. This approach allow to convert α-olefins via a one-pot procedure into *trans*-3,4-dialkyltetrahydrothiophenes with high selectivity in the yields of 65–80% through 1-ethyl-*trans*-3,4-dialkylaluminacyclopentanes (32) preliminary obtained [127]. Using the said method tri- and tetracyclic tetrahydrothiophenes have been synthesized from norbornenes. The reaction of generated *in situ* substituted aluminacyclopentanes with alkyl(phenyl)dichlorophosphines was recognized as the convenient rout to transform α-olefins into phospholanes (47) [128] (Scheme 42).

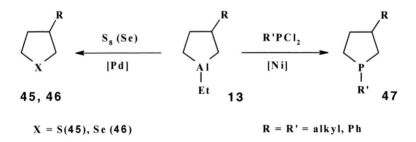

X = S(45), Se (46)   R = R' = alkyl, Ph

Scheme 42.

Among known, the method for the synthesis of cyclopentanols (48) and 1-hydroxycyclopentanecarboxylates (49), which based on consecutive cycloalumination of α-olefins with trialkylalanes in the presence of Cp_2ZrCl_2 catalyst via subsequent interaction between generated *in situ* aluminacyclopentanes and alkyl carboxylates in the presence of catalytic amounts of Cu, Ni, Pd salts or complexes is considered to be the most attractive [128–130] (Scheme 43).

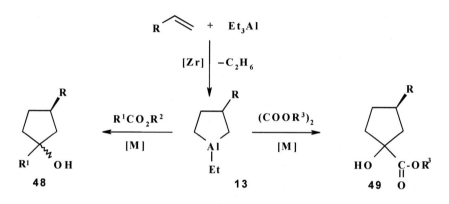

[M] = CuCl, CuI, Ni(acac)$_2$-Ph$_3$P

Scheme 43.

The elaborated on the basis of cyclic OAC method has been successfully applied to the synthesis of second and tertiary alcohols by the interaction between generated *in situ* 1-ethyl-3-alkylaluminacyclopentanes (13) and aldehydes or ketones at 20 °C in the presence of catalytic amounts of Cu salts. Together with OAC(13) 1-ethyl-*trans*-3,4-dialkylaluminacyclopentanes (32), polycyclic OAC (27, 28), 1-ethyl-2,3-fullero[60]aluminacyclopentanes (29) have been implicated in this reaction to synthesize different alcohols including carbocyclic (50) and fullerene-containing (51) ones [131] (Scheme 44).

Catalyzed by Cp_2ZrCl_2 cycloalumination of 3-methyloct-7-en-1-ol with Et$_3$Al followed by hydrolysis of generated *in situ* ACP gave rise to alcohol (52), which can be used as a synthon in synthesis of flour beetles, Tribolium confusum and Tribolium costaneum (53) [132] (Scheme 45).

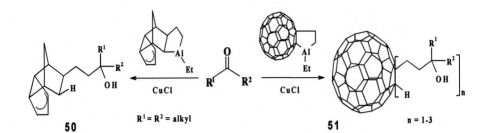

Scheme 44.

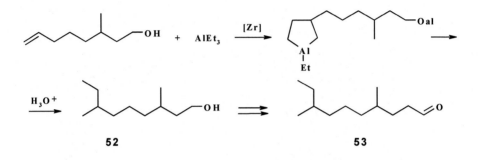

Scheme 45.

The insect pheromones (pine sawfly Neodiprion sertifer, German (red) cockroach Blattella germanica) have been synthesized by reductive β-vinylation of α-olefins through the preliminary generated *in situ* aluminacyclopentanes [133]. The catalyzed by Cp_2ZrCl_2 cycloalumination reaction of unsaturated compounds were also used for the synthesis of isoprenoids [134] and butane-1,4-diols [135] (Scheme 46).

The examples under consideration give evidence that olefins can be easily converted to carbocyclic, heterocyclic and also functionally substituted acyclic compounds by one pot procedure through the stage of starting olefin preliminary cycloalumination. For a long time one could not implicate 1,1-disubstituted olefins in the above reaction.

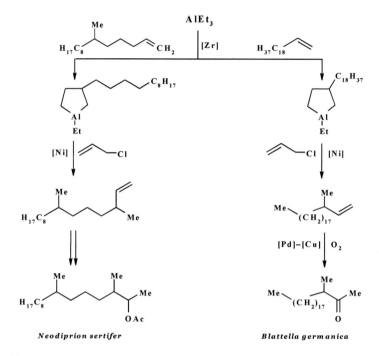

Scheme 46.

Comparatively recently the authors of the work [136] established that in contrast to the acyclic hydrocarbons with low active 1,1-disubstituted double bond methylenecycloalkanes can be implicated in the cycloalumination reaction. Cycloalumination of methylenecyclobutane with the excess of Et₃Al in the presence of 5 mole % Cp₂ZrCl₂ (4 h, pentane) resulted in 6-ethyl-6-aluminaspiro[3.4]octane (54) in more than 90% yield (Scheme 47).

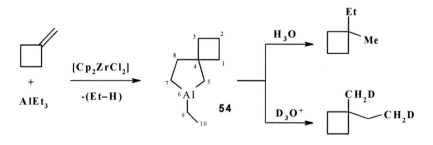

Scheme 47.

During these investigations the authors of the work [136] have performed catalytic cycloalumination of 3-methylene-*exo*-tricyclo[4.2.1.0^{2,5}]nonane (55), 3-methylene-*exo*-pentacyclo[5.4.0.0^{2,5}.0^{6,10}.0^{9,11}]undecane (56) and 9-methylene-*endo-exo*-tetracyclo[5.4.1.0^{2,6}.0^{8,11}]dodec-3(4)-ene (57a,b) under reaction conditions described above and obtained the appropriate alumina[3.4]octanes namely tricyclo[4.2.1.0^{2,5}]nonane-3-spiro(2'-ethyl-2'-aluminapentane) (58), pentacyclo[5.4.0.0^{2,5}.0^{6,10}.0^{9,11}]undecane-3-spiro(2'-ethyl-2'-aluminapentane) (59) and tetracyclo[4.5.1.0^{2,6}.0^{8,11}]dodec-3(4)-ene-9-spiro(2'-ethyl-2'-aluminapentane) (60a,b). Generated *in situ* aluminaspiro[3.4]octanes easily reacted with S_8 or Se giving rise to the corresponding spirotetrahydrothiophenes and selenophenes [137, 138] (Scheme 48).

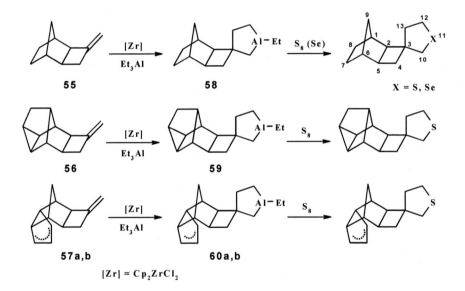

Scheme 48.

The catalytic cycloalumination reaction has been successfully used in the synthesis of spiro compounds. Thus, from 6-ethyl-6-aluminaspiro[3.4]octane (**54**) there have been synthesized spiro[3.3]heptane, 6-thiaspiro[3.4]octane, 6-spiro[3.4]octyl formiate, spiro[3.4]octanol, and 5-[1-(3-butenyl)cyclobutyl]pent-1-en [136, 138] (Scheme 49).

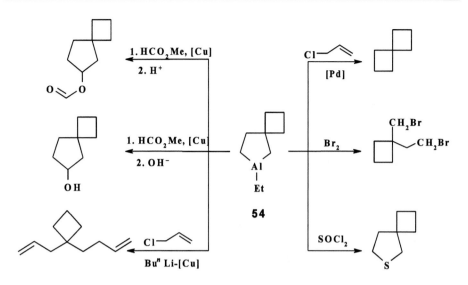

Scheme 49.

The contribution [41], in which the authors described the procedure for a synthesis of aryl substituted aluminacyclopropanes (61) via cycloalumination of arylolefins (sterene, *ortho-* and *para-*methyl sterenes), first appeared in 1997. The same work, additionally, demonstrated the synthesis of 1,4-diphenyl-1,3-butadiene with $EtAlCl_2$ in 65–85% yield in the presence of metallic Mg as an acceptor of halogenide ions and Cp_2TiCl_2 catalyst [139]. Small amounts (no more than 10 %) of aryl substituted aluminacyclopentanes have been also detected. The authors of these works assumed that generation from Cp_2TiCl_2, Mg and aryl olefins of titanacyclopropane intermediates, transmetalation of which with $EtAlCl_2$ led to aluminacyclopropanes (61) was the key stage of the reaction [140]. Under the same conditions fullerenes with annulated aluminacyclopropane fragments (62) have been obtained via the interaction between C_{60} and $EtAlCl_2$ (excess) in the presence of metallic Mg and Cp_2TiCl_2 catalyst in THF-toluene solution at 20 °C [141]. Proposed in the work [141] the Scheme for the formation of fullero[60]aluminacyclopropanes (62) through transmetalation of generated *in situ* titanacyclopropane intermediates appeared sufficiently reasonable since the procedure to synthesize $Cp_2Ti(\eta^2-C_{60})$ is described in the literature and its structure is also proven [142] (Scheme 50).

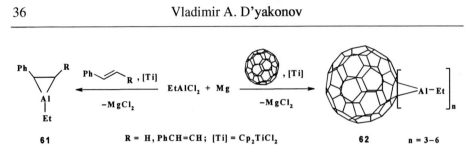

Scheme 50.

In 2001, 1,2-dienes together with 1,3-dienes have been successfully implicated in the cycloalumination reaction giving rise to the appropriate 2-alkylidenaluminacyclopentanes. As it was shown by the authors of the works [143] cycloalumination of 1,2-dienes with Et_3Al under the action of Cp_2ZrCl_2 (5 mole %, ~20 °C, 4 h) occurs in aliphatic (70–80 %), aromatic (75–77 %) solvents and also in methylene dichloride (92%) to yield 1-ethyl-2-alkylidenealuminacyclopentanes (63) (Scheme 51). In ethereal solvents (THF) or without the solvent the reaction proceeded to afford the product in low yield (less than 20%). In these experiments 1-ethyl-2-methylene-3-alkylaluminacyc lopentanes were shown to obtain in minor quantities (3–15%). Chemical transformations of cyclic OACs (63) led to corresponding olefins and alcohols containing Z-disubstituted double bonds.

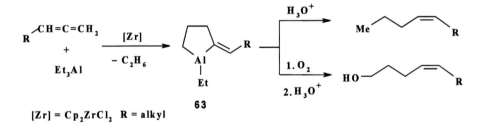

Scheme 51.

Cycloalumination of acylic allenes with $EtAlCl_2$ in the presence of Cp_2TiCl_2 (5 mole %, ~20 °C) and metallic Mg (acceptor of halogenide ions) was found to produce 1-ethyl-2-methylene-3-alkylaluminacyclopropanes (64) and 1-ethyl-2,5-dialkylidenealuminacyclopentanes (65) in total 80% yield [42]. The replacement of $EtAlCl_2$ by Et_2AlCl allowed to synthesize the corresponding 1,2- (66) and 1,4-dialuminum (67) compounds [42] (Scheme 52).

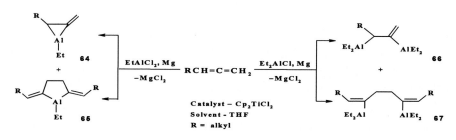

Scheme 52.

The authors of the work [144, 145] have succeeded in expanding application boundaries of the cycloalumination reaction on the example of cyclonona-1,2-diene using Et₃Al or EtAlCl₂ and Cp₂ZrCl₂ as a catalyst. Thus, the interaction between cyclonona-1,2-diene and Et₃Al (excess) under chosen conditions (5 mole %, , 4 h) was shown to afford 10-ethyl-10-aluminabicyclo[7.3.0^{1,9}]dodec-8-ene (68) with high regioselectivity (> 95%) in the more than 85% yield [145]. Cycloalumination of cyclonona-1,2-diene occurred in aliphatic (hexane, cyclohexane), aromatic (benzol) solvents, and also in CH₂Cl₂ for 4−5 hours (Scheme 53). In ethereal solvents (THF, diethyl ether) or without the solvent the reaction proceeded with low yield as a result of polymerization of the starting allene.

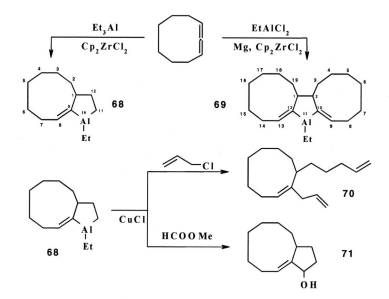

Scheme 53.

The reaction of OAC (68) with allyl chloride or methyl formate [145] in the presence of CuCl (10 mole %) allowed synthesizing olefin (70) or cyclopentanol (71) by one pot procedure. The intermolecular cycloalumination of cyclonona-1,2-diene with $EtAlCl_2$ in the presence of metallic Mg and Cp_2ZrCl_2 catalyst in THF gave rise to 11-ethyl-11-aluminatricyclo[$10.7.0^{1,12}.0^{2,10}$]nonadeca-9,12-diene (69) (Scheme 53).

An interesting finding in the construction of new aluminacarbocycles with a given structure is a combine cycloalumination of cyclic 1,2-dienes **and** α-olefins, terminal allenes, disubstituted acetylenes using alkylhalogenalanes in the presence of Mg catalyzed by Zr and Ti complexes [146, 147]. Using this approach the methods to obtain in high yields previously hard-to-reach bicyclic aluminacyclopentanes (73) and aluminacyclopentenes (72) including alkyl-substituted (74, 75) have been developed (Scheme 54).

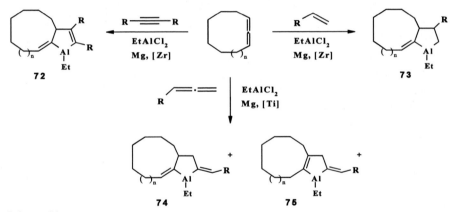

Scheme 54.

The aforesaid procedures to synthesize novel classes of cyclic organoaluminum compoundes give evidence that catalytic cycloalumination reaction of unsaturated compounds is of general nature and allows to convert olefins and dienes into three-, five-, and seven-membered OAC having very high reactivity.

Such reaction shows great synthetic potential and can be successfully used in organic and organometallic chemistry.

4.2. CATALYTIC CYCLOALUMNATION OF ACETYLENES TO ALUMINACYCLOPENTENES AND ALUMINACYCLOPENTADIENES

During further investigations of the catalytic cycloalumination reaction of unsaturated compounds Professor Dzhemilev in 1990 first evidenced for the involvement of acetylene in the said reaction that would, according to the authors [39], allow carrying out the synthesis of aluminacyclopentenes and aluminacyclopentadienes [148].

As a result, 1-ethyl-2,3-dialkyl(aryl)-2-aluminacyclopent-2-enes (76) have been obtained in 75−90% yield by the reaction of disubstituted acetylene with Et_3Al in the presence of Cp_2ZrCl_2 catalyst at 20 °C [43, 149] (Scheme 55). The structure of (76) was characterized by ^{13}C NMR spectroscopy [150] and also by chemical transformations. Thus, alkylation of (76) with dimethylsulfate or diethylsulfate occurs at the double bond in the ring to afford gomoallylic OACs (77). The latters were in turn underwent intramolecular carboalumination allowing 1,1-disubstituted cyclopropane (78) after re-alkylation [151].

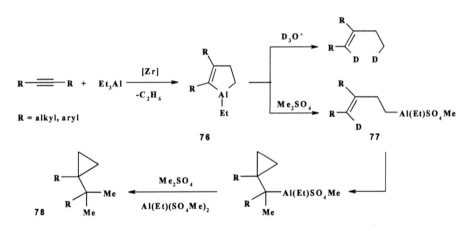

Scheme 55.

Six years after the first communication of Prof. U.M. Dzhemilev and coworkers [148, 149] on the synthesis of aluminacyclopentenes by cycloalumination of acetylene using $AlEt_3$ in the presence of the Cp_2ZrCl_2 catalyst, Prof. Ei-ichi Negishi and coworkers [96] has been reported the performance of intramolecular cycloalumination of α,ω-enynes to appropriate bicyclic aluminacyclopentenes (109) (Scheme 56).

It is regrettable to note that both at that time [152], and later, in all subsequent reports of E. Negishi and coworkers [26], the researches on the synthesis and transformations of aluminacyclopentenes as well as the mechanistic studies of the cycloalumination reaction are presented as the contuation of their own unpublished work.

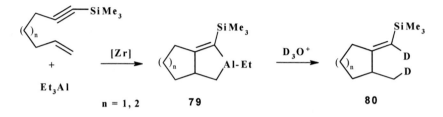

Scheme 56.

Cycloalumination of 1,4-enynes with Et_3Al under the action of Cp_2ZrCl_2 (5 mole %, ~20 °C, 8 h, hexane) led to regioisomeric mixture of 2,3-disubstituted aluminacyclopent-2-enes (81) with retention of the original double bond at the allylic position [43]. One could incorporate the latter into the cycloalumination reaction only under the interaction between 1,4-enynes and four-fold excess of Et_3Al in the presence of Cp_2ZrCl_2 catalyst (10−15 mole % towards the starting 1,4-enyne, 20−21 °C, 8−10 h). As a result, regioisomeric (~1:1 ratio) (aluminacyclopent-3-ylmethyl)-aluminacyclopent-2-enes (82) have been identified as 1,1-dialkylsubstituted cyclopropanes (83) through the transformation under the effect of Me_2SO_4 [44] (Scheme 57).

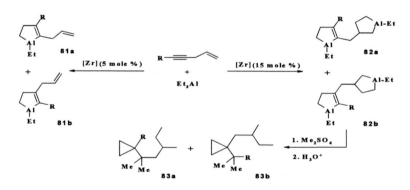

Scheme 57.

Aluminacyclopentenes (79) obtained by the reaction of 1,6- or 1,7-enynes with Et_3Al in the presence of Cp_2ZrCl_2 were found to interact with CO_2 (0 °C, 1 atm) or $ClCH_2OCH_3$ (23 °C) yielding bicyclic cyclopentanones (84) or vinylcyclopropanes (85) [152] (Scheme 58).

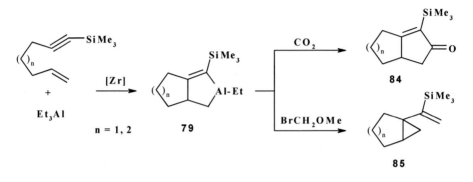

Scheme 58.

Modified method for a synthesis of 2,3-dialkyl(phenyl)aluminacyclopentenes by cycloalumination of disubstituted acetylenes with $EtAlCl_2$ in the presence of ethylene generated from 1,2-dichloroethane and activated magnesuim under the action of Cp_2TiCl_2 catalyst has been considered in the work [153] (Scheme 59). The reaction was found to proceed at ambient temperature in tetrahydrofurane. Together with aluminacyclopentenes (76) the small amounts of aluminacyclopropenes (86) and aluminacyclopentadienes (87) have been detected. The yields and ratios of the latters were shown to depend upon the nature of substituents in the starting disubstituted acetylene.

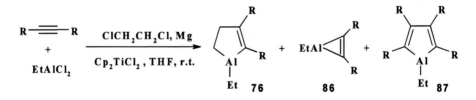

Scheme 59.

Based on the achievement data above the authors of the work [154] have elaborated the methodology for the synthesis of tetrasubstituted aluminacyclopent-2-enes by combined cycloalumination of acetylenes and olefins. Thus, combined cycloalumination of disubstituted acetylene and α-olefin with EtAlCl$_2$ in the presence of Cp$_2$ZrCl$_2$ catalyst and metallic magnesium (THF, 20−22 °C, 8 h) was found to provide expected 1,2,3,4-tetraalkylaluminacyclopent-2-ene (88) as a major product together with 1,2,3,4,5-pentaalkylaluminacyclopenta-2,4-diene (87) and 1-ethyl-*trans*-3,4-dialkylaluminacyclopentane (32) in minor amounts. The combined product yield reached over 80% (Scheme 60).

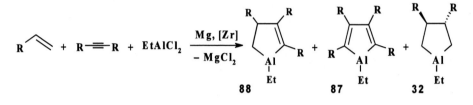

Scheme 60.

One could alter the direction of the chemical reaction preferably to the formation of aluminacyclopent-2-ene (88) under slow component addition conditions *viz.*, **slow addition (during 6 h) of** α-olefin and EtAlCl$_2$ in THF to the toluene solution of Cp$_2$ZrCl$_2$ catalyst containing acetylene and metallic Mg.

The same procedure as described above has been developed under cycloalumination of disubstituted acetylenes with higher trialkyl alanes under the action of Cp$_2$ZrCl$_2$ [154]. As a result, 1,2,3,4-tetrasubstituted aluminacyclopent-2-enes (89) have been obtained in 50−55 % yield (Scheme 61). Together with (89) in the same experiments the minor amounts (10−15 %) of 1,2,3,4,5-tetraalkylsubstituted aluminacyclopenta-2,4-dienes (90) and 1-alkyl-*trans*-3,4-dialkylsubstituted aluminacyclopentanes (91) have been detected in the reaction mixture.

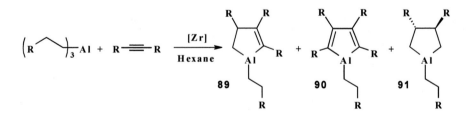

Scheme 61.

In 1992, the new preparative procedure to synthesize aluminacyclopenta-2,4-dienes (**87**) based on the cycloalumination reaction of disubstituted acetylenes with RAlCl$_2$ (R = Et, BuO, Et$_2$N, Cl) affected by Cp$_2$ZrCl$_2$ catalyst and widely used in cycloalumination of acetylenes with the aid of Et$_3$Al has been proposed [43, 44, 149]. The authors of the work [44] proceeded from the assumption that reduction of Cp$_2$ZrCl$_2$ with Mg in the presence of disubstituted acetylenes gave rise to zirconacyclopentadienes (92) [155], which then transmetallated with EtAlCl$_2$ to aluminacyclopentadienes (87) (Scheme 62).

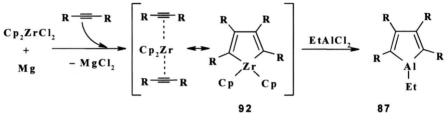

Scheme 62.

One should note that during intermolecular cycloalumination of phenylcetylene with EtAlCl$_2$ in the presence of metallic Mg affected by Cp$_2$ZrCl$_2$ under the chosen reaction conditions [44] simultaneous incorporation of the double and triple bonds occurred leading selectively to tricyclic dialuminum compound (93) (Scheme 63).

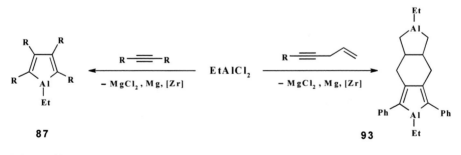

Scheme 63.

Together with acyclic acetylenes one has also succeeded in catalytic cycloalumination of cyclic acetylenes. Thus, cycloalumination of octyne and cyclododecyne with EtAlCl$_2$ or AlCl$_3$ in the presence of metallic Mg affected by Cp$_2$ZrCl$_2$ catalyst under reaction conditions (r.t., 6 h, THF) opens a convenient synthetic rout towards tricyclic aluminacyclopentadienes (94) in the yields of more than 70% [156] according the following Scheme 64:

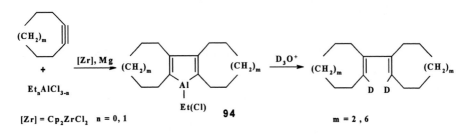

Scheme 64.

Based on the results above the combined cycloalumination reaction of cycloalkynes and acyclic disubstituted acetylenes has been realized to obtain the new types of bicyclic aluminacyclopentadienes. Thus, the combined intermolecular cycloalumination of cyclooctyne and hex-3-yne mixture with the aid of EtAlCl$_2$ in the presence of Cp$_2$ZrCl$_2$ catalyst and metallic Mg (20 °C, 6 h, THF) has been successfully performed giving rise to 9,10,11-triethyl-9-aluminabicyclo[2.6.0^{1,8}]undeca-1,10-diene (95) in more than 55% yield (Scheme 65). Together with the target OAC (95) the small amounts (less than 10%) of pentaethylaluminacyclopenta-2,4-diene (96) have been observed as the initial hexyne cycloalumination product [156].

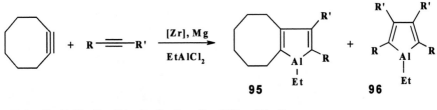

[Zr] = Cp$_2$ZrCl$_2$; R = R' = Et, Pr, Bu ; R = SiMe$_3$, R' = Bu

Scheme 65.

Based on the achievement data of M.E. Vol'pin and coworkers [157−159], *viz.,* the ability of low valence titanium complexes to form titanacyclopropenes through the coordination of acetylenes, the authors of the works [45] in 1997 have succeeded in synthesizing of aluminacyclopropenes (97) via catalytic cycloalumination of disubstituted acetylenes with EtAlCl$_2$. Together with target OAC (97) the small amounts of substituted aluminacyclopentadienes and substituted benzenes have been obtained (Scheme 66). The structure of aluminacyclopropenes was determined by spectral methods.

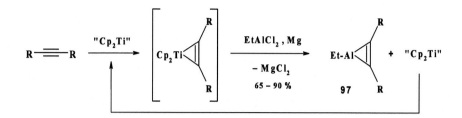

Scheme 66.

Cyclometalation of acetylenes with bulky substituents, for example, 1-phenyl-2(trimethylsilyl)-acetylene with ElAlCl$_2$ under reaction conditions [ElAlCl$_2$: 1-phenyl-2(trimethylsilyl)acetylene : Mg : Cp$_2$TiCl$_2$ = 200:100:100:5, r.t., 8 h, THF] was shown to afford 1-ethyl-2-phenyl-3-(trimethylsilyl)aluminacycloprop-2-ene (**98**) in the yield of no more than 15%. However with the increase in duration of the reaction to 72 h and the use of 10 mole % Cp$_2$TiCl$_2$ the yield of target (**98**) also increased to 55% [160] (Scheme 67). .

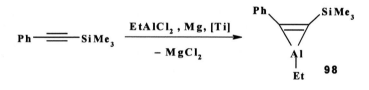

Scheme 67.

The replacement of EtAlCl by Et$_2$AlCl changed the direction of the catalytic cycloalumination reaction giving rise to 1,2-dialuminio ethenes. Tolane entered the said reaction under chosen conditions (Cp$_2$TiCl$_2$, 10 mole %, r.t., 8 h, THF) more selectively to obtain 1,2-diphenyl-1,2-bis(diethylaluminio)ethene (**99**) in 70% total yield (Scheme 68). In an analogous fashion, substituted 1,2-dialkyl acetylenes underwent cycloalumination to 1,2-dialuminioethenes but the yield in this reaction did not exceed 50 % [161].

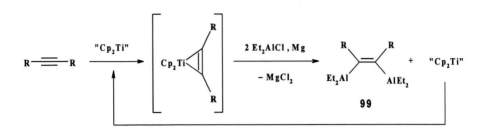

Scheme 68.

The authors of the work [88] have first announced that together with Ti- and Zr-containing complex catalysts, which were employed in catalytic cyclometalation of unsaturated compounds, Co phosphine complexes could provide appropriate cyclic OACs in the olefin, allene and also acetylene cycloalumination reactions with the aid of trialkyl and alkyl halogenalanes (Scheme 69).

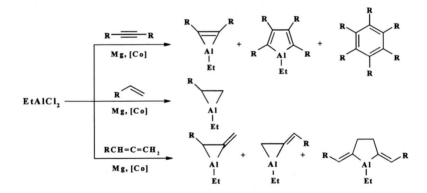

Scheme 69.

4.3. CYCLOMAGNESIATION OF OLEFINS AND 1,2-DIENES TO MAGNESACYCLOPENTANES AFFECTED BY ZR- AND TI-CATALYSTS

The first report on catalytic cyclomagnesiation of olefins with RMgX or R_2Mg to the corresponding magnasacyclopentanes has appeared in 1989. The authors of the work [162] have shown that the reaction of styrene with R_2Mg under the effect of Cp_2ZrCl_2 catalyst led under mild conditions (20 $^\circ$C, Et_2O/THF)

to a 1:3 mixture of magnesacyclopentanes (100) and (101) in a total 80% yield [163] (Scheme 70). Such mixture of magnesacyclopentanes is formed as a result of catalytic cyclomagnesiation of *o*-, *m*-, *p*-methyl or *m-tert*-butyl styrenes [164, 165].

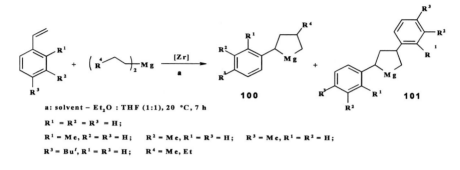

a: solvent – Et$_2$O : THF (1:1), 20 °C, 7 h

R^1 = R^2 = R^3 = H;

R^1 = Me, R^2 = R^3 = H;   R^2 = Me, R^1 = R^3 = H;   R^3 = Me, R^1 = R^2 = H;

R^3 = Bu^t, R^1 = R^2 = H;   R^4 = Me, Et

Scheme 70.

Unlike styrene, cyclomagnesiation of norbornenes such as bicyclo[2.2.1]hept-2-ene and spiro{bicyclo[2.2.1]hept-2-ene-7,1'-cyclopropane} with Pr$_2^n$Mg or Bu$_2^n$Mg under the action of catalytic amounts of Cp$_2$ZrCl$_2$ (3 mole %) in Et$_2$O/THF solution (22 °C, 8 h) led to diastereomeric pure tri-, tetra-, penta-, and heptacyclic OMCs, *viz.*, *exo,exo*-5-alkyl-3-magnesatricyclo[5.2.1.0^{2,6}]decane, *exo,exo*-5-alkyl-3-magnesaspiro{tri-cyclo[5.2.1.0^{2,6}]decane-10,1'-cyclopropane}, *exo,exo*-9-magnesapentacyclo[9.2.1.1^{4,7}.0^{2,10}.0^{3,8}]pentadecane, and *exo,exo*-9-magnesaspiro{pentacyclo[9.2.1.1^{4,7}.0^{2,10}.0^{3,8}]pentadecane-14,1'(15,1')-dicyclopropane in 80 to 95 % yield [166] (Scheme 71).

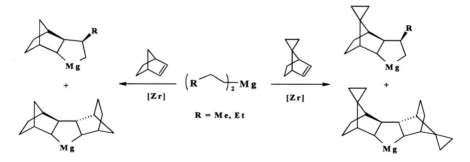

R = Me, Et

Scheme 71.

In contradistinction to R_2Mg (R = Pr, Bu, Hex, octyl) the organomagnesium reagents EtMgX and Et_2Mg were found to react with α-olefins in the presence of catalytic amounts of Cp_2ZrCl_2 to yield the ethylmagnesiation or cyclomagnesiation products according to the reaction conditions. In this case one can regulate the direction of the reaction by controlling the solvent nature, reaction temperature and the ratio of initial reagents. Thus, the interaction between EtMgX and $RCH_2CH=CH_2$ (2:1 ratio) was shown to afford the ethylmagnesiation products (102) and (103) (95:5 ratio) at ambient temperature in THF. If the reaction was carried out at a ratio Mg:olefin equal to 4:1 in diethyl ether at 0 °C using Et_2Mg instead of EtMgX magnesacyclopentanes (and/or 1,4-dimagnesium compounds) (103) were predominantly (~85%) obtained [31] (Scheme 72).

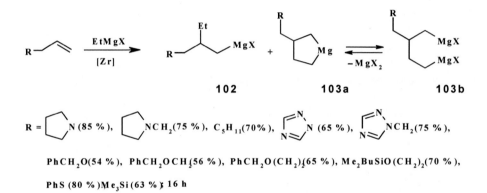

102   **103a**   **103b**

R = ![ring]N (85 %), ![ring]NCH₂(75 %), C_5H_{11}(70%), ![ring]N (65 %), ![ring]NCH₂(75 %),

PhCH₂O(54 %), PhCH₂OCH(56 %), PhCH₂O(CH₂)(65 %), Me₂BuSiO(CH₂)₂(70 %),

PhS (80 %)Me₃Si(63 %) 16 h

Scheme 72.

Mechanism of the Cp_2ZrCl_2 catalyzed ethylmagnesiation and cyclomagnesiation reactions of olefins with nonactivated double bond have been discussed by several groups of researchers at one time [31, 163, 167−169].

Cyclomagnesiation of α-olefins (oct-1-ene, allyl benzene, styrene, *endo*-dicyclopentadiene) with EtMgR (R = Br, Et) affected by Ti-containing catalysts has been reported for the first time in the work [170]. The interaction between EtMgBr or Et_2Mg and olefins in the presence of Cp_2TiCl_2 catalyst unlike those affected by Zr-containing ones was shown to provide the cyclo-, carbo- or hydromagnesiation products depending upon the nature of olefin reactant. Thus, under the interaction between *endo*-dicyclopentadiene (DCPD) and EtMgBr (20 °C, 50 h, THF, DCPD: EtMgBr: [Ti] = 1:2:0,05) primarily carbomagnesiation of the norbornene double bond occurred giving regio isomeric (~1:1 ratio) ethylmagnesiation products (105) in ~70 % yield (Scheme 73). But in the presence of chemically activated Mg [171] one could observe the alteration in reaction chemoselectivity and the formation together with (105) of the cyclomagnesiation (104) and hydromagnesiation (106) products in the (104):(105):(106) ratio equal ~ 5:2:2 with retention of *endo*-configuration of cyclopentene moiety. The chemically activated Mg took part in reduction of Cp$_2$TiCl$_2$ to "Cp$_2$Ti" being responsible for the generation of titanacyclopentane intermediates [172]. The cyclomagnesiation and 1,2-ethylmagnesiation reactions are characterized by high stereoselectivity, and addition of organomagnesium reagent to the norbornene double bond in every experiment occurred strongly from the *exo*-side of the molecule.

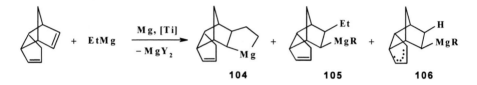

104   **105**   **106**

Scheme 73.

One could perform intramolecular cyclization of α,ω-non-conjugated dienes using Grignard reagents and Cp$_2$ZrCl$_2$ as a catalyst. Thus, the Cp$_2$ZrCl$_2$ catalyzed reaction of non-conjugated α,ω-dienes with BuMgX or Bu$_2$Mg in ethereal solvents led to 1,2-di(halomagnesamethyl)-substituted carbocycles (107) or 1-(halomagnesamethyl)-2-methyl-substituted carbocycles (108) [169, 173−175] (Scheme 74). Yields, selectivity and stereochemistry of the obtained carbocycles (107) and (108) were shown to depend upon the structure of α,ω-dienes, the nature of organomagnesium reagents as well as reaction conditions and ratio of initial reagents.

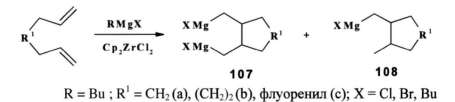

107   **108**

R = Bu ; R^1 = CH$_2$ (a), (CH$_2$)$_2$ (b), флуоренил (c); X = Cl, Br, Bu

Scheme 74.

Stereochemistry of carbo- and cyclomagnesiation of α-olefins and non-conjugated dienes under interaction with R$_2$Mg or RMgX in the presence of chiral

Zr complexes have been discribed in the works [176–178]. As shown [178], the interaction between α,ω-diolefins and n-BuMgR (R = n-Bu,Cl) affected by Zr complexes was stated to afford intermediate diastereomers (109) and (110), subsequent transformations of which under reaction conditions led to *cis*- and *trans*-1,4-dimagnesium compounds (111) and (112) (Scheme 75).

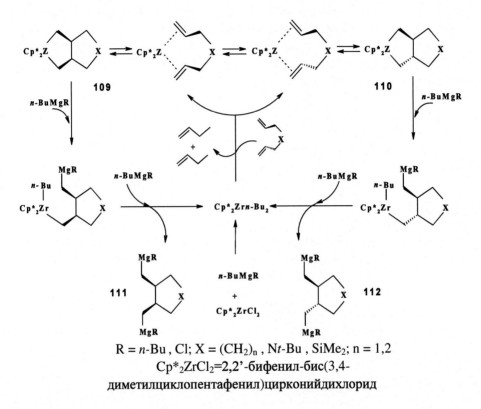

R = n-Bu , Cl; X = $(CH_2)_n$, Nt-Bu , SiMe$_2$; n = 1,2
Cp*$_2$ZrCl$_2$=2,2'-бифенил-бис(3,4-диметилциклопентафенил)цирконийдихлорид

Scheme 75.

The carbocyclization reaction of α,ω-diolefins with RMgX under the action of the catalysts based on Zr complexes is considered as effective and promising method for a synthesis of the hydroindane derivatives (113) including optically active ones [179–189], according to the following Scheme 76:

The authors of the work [190] for the first time have succeeded in involving 1,2-dienes into the cyclomagnesiation reaction. The carbomagnesiation products (114-116) were manifested to form in the presence of Cp$_2$ZrCl$_2$ catalyst at ambient temperature in THF. The decrease in temperature to 0 °C and Et$_2$O media provided the formation of magnesacyclopentanes (117-119) (Scheme 77).

Cyclomagnesiation of 1,2-dienes assisted by two-fold excess of EtMgBr in the presence of chemically activated Mg and catalytic amounts (5 mole %) of Cp_2TiCl_2 under reaction conditions (THF, r.t., 8 h) resulted in 2,5-dialkylidenemagnesacyclopentanes and 1,4-dimagnesium compounds in the (120a) : (120b) ratio equal to ~ 1 : 1 according to ^{13}C NMR spectral data [190] (Scheme 78).

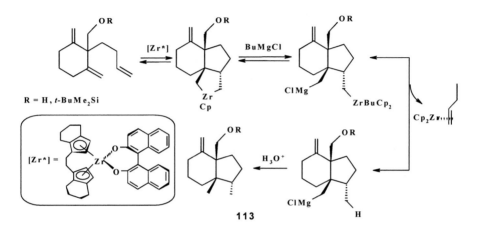

Scheme 76.

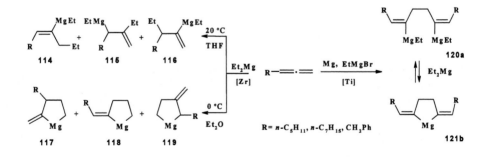

Scheme 77.

Apparently, reducing of Cp_2TiCl_2 to «Cp_2Ti», generation of intermediate 2,5-dialkylidenetitanacyclopentanes (121) and subsequent transmetalation of the latters with EtMgX to target products (120) appeared as the most probable intermediate steps, which could explain the formation of unsaturated OMCs (120a) and (120b). It was shown the possibility to realize one pot conversion of

(120) to Z-diolefins (122, 123) with the aid of organic halogenides under the action of cuprous salts [191].

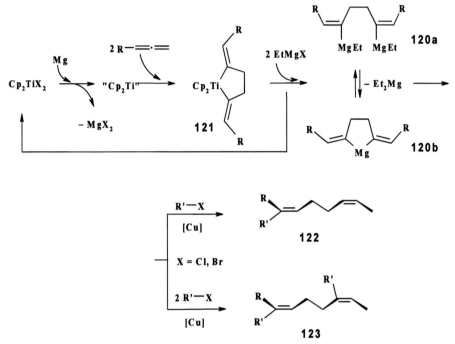

Scheme 78.

For the first time the authors of the work [192] have realized catalytic cyclo- and ethylmagnesiation of cyclonona-1,2-diene to develop these prospective investigations in the presence of Cp$_2$ZrCl$_2$ catalyst (5 mole %) in THF or Et$_2$O using EtMgR (R = Et, Hlg) to afford 10-magnesabicyclo[7.3.0^{1,9}]dodec-8-ene (124) or 3-ethylcyclonon-1-en-2-yl magnesium ethyl (125) depending upon experimental conditions (Scheme 79). The interaction between Et$_2$Mg and cyclonona-1,2-diene in the presence of 5 mole % Cp$_2$ZrCl$_2$ in diethyl ether at 0 °C was found to produce OMCs (124) and (125) (91:9 ratio) in 79% combined yield. The action of methyl formate on magnesabicyclane (124) in the presence of CuCl catalyst (10 mole %) gave rise to 10-hydroxybicyclo[7.3.0^{1,9}]dodec-8-ene (126) in 67% yield. The reaction of EtMgBr with cyclonona-1,2-diene in THF at ambient temperature resulted in the predominant formation of the carbomagnesiation product (125). In this case deuterolysis of the reaction mixture led to mono- (127) and dideuterized (128) hydrocarbons (95:5 ratio) in 56% combined yield.

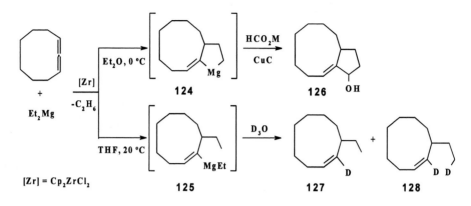

Scheme 79.

The same authors of the work [193] have established that cyclonona-1,2-dienes easily entered the reaction with EtMgBr in the presence of metallic Mg (acceptor of halogenide ions) and 5 mole % Cp_2TiCl_2 catalyst (Et_2O, 4 h, r.t.) to obtain 11-magnesatricyclo[$10.7.0^{1,12}.0^{2,11}$]nonatrideca-3(4),19-diene (and/or 1,4-dimagnesium compound) (129). Deuterolysis of the latter led to 2-deutero-3-(2-deutero-2-cyclononenyl)-1-cyclononene (130) in 85% yields. The OMC (129) also reacts with elemental sulfur S_8 to provide thiophane (131), which quantatively isomerizes to thiophene (132) while heating up to 130–140 °C. When dry CO_2 was bubbled through the reaction mixture containing OMC (129), the latter then transformed to unsaturated tricyclic ketone (134) in 75% yield. The CuCl catalyst initiated intramolecular cyclization of (129) to produce (10R,11S)-tricyclo[$9.7.0^{1,11}.0^{2,10}$]octadeca-2(3),18-diene (133) in 68% yield (Scheme 80).

The newly developed reaction [193] allowed to involve cyclic and acyclic 1,2-dienes in the combined intermolecular cyclomagnesiation with Grignard reagents and Cp_2TiCl_2 catalyst to obtain new types of bicyclic alkylidenemagnesacyclopentanes (and/or 1,4-dimagnesium compounds). Intermolecular cyclomagnesiation of cyclonona-1,2-diene and hepta-1,2-diene with EtMgBr (excess) in the presence of chemically active Mg and Cp_2TiCl_2 catalyst was shown to afford 11-pentalidene-12-magnesabicyclo[$7.3.0^{1,2}$]dodec-2(3)-ene (135) under reaction conditions (Et_2O, 4 h, 20 °C) in 88% yield (Scheme 81).

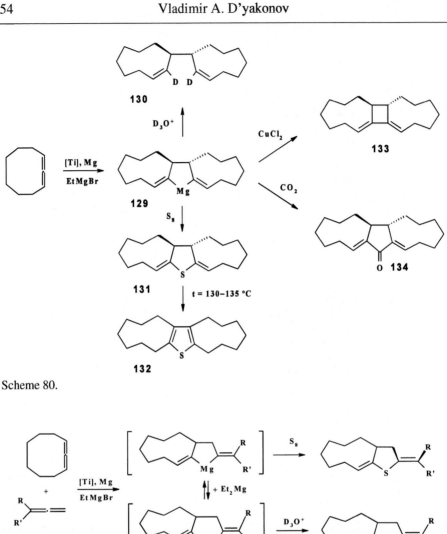

Scheme 80.

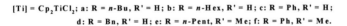

135a–f

[Ti] = Cp$_2$TiCl$_2$; a: R = n-Bu, R' = H; b: R = n-Hex, R' = H; c: R = Ph, R' = H;
d: R = Bn, R' = H; e: R = n-Pent, R' = Me; f: R = Ph, R' = Me.

Scheme 81.

The formation of the cyclonona-1,2-diene and hepta-1,2-diene homocyclomagnesiation products (~1:1 ratio) has been observed as a minor component in the yield of no more than 8–10%.

4.4. CATALYTIC CYCLOMAGNESIATION OF ACETYLENES

The information concerned the catalytic cyclomagnesiation reactions of acetylenes [194] in comparison with the hydro- and carbomagnesiation ones has not been described in scientific literature up to nowadays.

As shown in late 2006 and early 2007, the reaction of disubstituted acetylenes with BuMgBr in Et$_2$O in the presence of Cp$_2$ZrCl$_2$ catalyst (10 mole %) under mild conditions (r.t., 2 h) gave rise to tetrasubstituted magnesacyclopentadienes (136) in 50% yield [195] (Scheme 82). The replacement of *n*-BuMgBr by *n*-BuMgCl did not noticeably influence the yield of the target magnesacyclopentadiene, but the yield of (136) did not exceed 15% while using THF as a solvent.

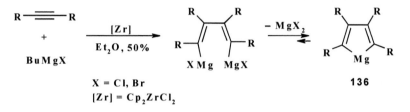

Scheme 82.

The synthesis of 2,3-dialkyl-5-alkylidenemagnesacyclopent-2-enes (137) has been performed via intermolecular cyclomagnesiation of disubstituted acetylene and allene in equimolar amounts with *n*-BuMgX (X = Cl, Br) affected by Cp$_2$ZrCl$_2$ in Et$_2$O under optimized reaction conditions [195] (Scheme 83). In these experiments together with target magnesacyclopentenes (137) the corresponding magnesacyclopentadienes (136) were shown to obtain as a minor product (<15%).

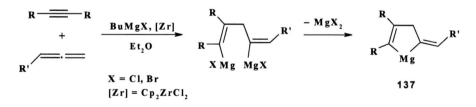

Scheme 83.

The reaction of generated *in situ* magnesacyclopentadienes (136) with elemental sulfur (S_2Cl_2, thionyl chloride) under mild conditions led to tetrasubstituted thiophenes (138) [138] (Scheme 84).

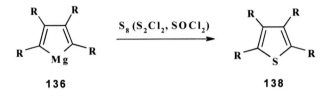

Scheme 84.

During investigations of the cycloalkyne cyclomagnesiation reaction using Gringard reagents in the presence of Cp_2ZrCl_2 catalyst the authors of the work [196] found that chemoselectivity of the reaction significantly depends upon temperature and solvent. Thus, the above reaction at 0 °C for 8 h gave rise to magnesacyclopentadienes (141) in 60% yield and with selectivity more than 98%. Together with OMCs (141) one could observe the noncatalytic carbomagnesiation products (139) and (140) if the reaction were conducting at 40 °C in diethyl or diisopropyl ether (Scheme 85).

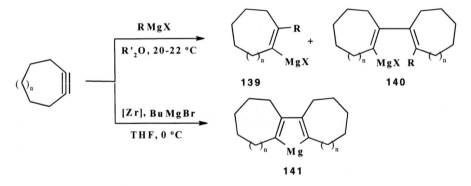

$[Zr] = Cp_2ZrCl_2$; n = 2, 6; R' = Et, *i*-Pr; R = Bu, Et; X = Bu, Et, Br.

Scheme 85.

To confirm the structure of the obtained tricyclic magnesacyclopentadienes (141) a series of transformations involving elemental sulfur, α,α'-dibromo-*o*-xylene and methyl iodide have been carried out to produce 2-thiatricyclo [9.6.0^{1,11}.0^{3,10}]heptadeca-1(11),3(10)-diene (142), 3,4-benzotricyclo

$[12.6.0^{1,14}.0^{6,13}]$eicosa-1(14),6(13)-diene (143) and 2,2'-dimethyl-1,1'-bi(cyclooct-1-ene-1-yl) (144) [196] (Scheme 86).

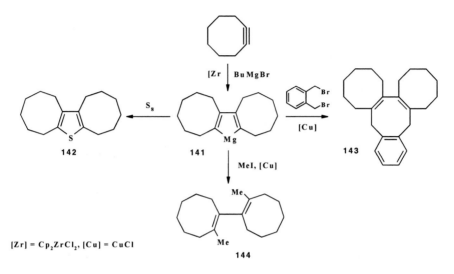

Scheme 86.

High reactivity of Mg–C bonds demonstrated by magnesacyclopentenes and magnesacyclopentadienes makes this class of metallacarbocycles exclusively promising for a synthesis of wide assortment of organic and organometallic compounds.

The above literature data give evidence that the catalytic cycloalumination and cyclomagnesiation reactions of olefins, acetylene, dienes and allenes resulted in the development of the universal one pot procedures appeared due to Professor Dzhemilev as synthetic tools for the new classes of three-membered, five-membered, and macrocyclic as well as 1,2- and 1,4-acyclic organoaluminum and organomagnesium compounds with high synthetic potential.

In the nearest future, one can expect the continued flow of new nontrivial results and contribution to the further development of these areas of research.

Chapter 5

ON MECHANISM OF OLEFIN ETHYLMAGNESIATION AND CYCLOALUMINATION CATALYZED BY ZR COMPLEXES

There are a large number of published works in world literature dealing with the study of the olefin polymerization mechanism with the use of zirconium catalysis and, to a lesser extent, with studies on the mechanism of organic and organometallic reactions involving zirconium complex as catalysts.

After the communication by Dzhemilev and coworkers in 1983 on the possibility of the catalytic ethylmagnesiation reaction of olefins with nonactivated double bond and at a later date the cyclometalation reaction of unsaturated compounds mediated by alkyl and halogenalkyl derivatives of magnesium and aluminum in the presence of the Cp_2ZrCl_2 catalyst a large number of reports devoted to the establishment of mechanisms of these reactions have appeared in the literature.

In addition to the above team, such famous scientists as A.H. Hoveyda, E. Negishi, R.J. Witby, T. Takahashi, R.M. Waymouth, K.S. Knight and many other professionals working in the area of zirconium catalysis have addressed this complicated and interesting problem. Their research are closely intertwined, complementing each other, raise new questions and then resolve them, and sometimes come into collision with existing experimental facts.

This chapter will be reviewed published results on the mechanistic study of the ethylmagnesiation and cycloalumination reactions catalyzed by Zr complexes.

In 1991 E. Negishi and T. Takahashi proposed the mechanistic scheme for the olefin ethylmagnesiation reaction assisted by EtMgBr and Cp_2ZrCl_2 as a catalyst. The sequence of transformations, in accordance with the suggested catalytic cycle, included the generation of zirconacyclopentane intermediate (145), which under the effect of EtMgBr through bimetallic complex (146) was converted to the target OMC (102). The formation of ethylene-zirconocene complex and regeneration of the intermediate (145) completed the proposed catalytic cycle (Scheme 87) [167, 197–199].

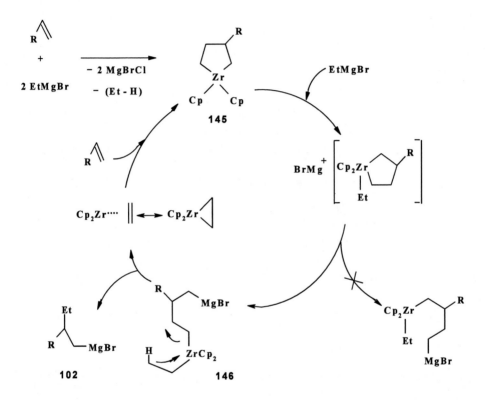

Scheme 87.

Similar conclusions about the structure of the key intermediates were made by A.H. Hoveyda with coworkers [66] during the study of the interaction between EtMgCl and cyclic allylic ethers including enantioselective addition of the original unsaturated compound to complex (147)with the formation of zirconacyclopentane intermediate (148). Regiocontrolled cleavage of zirconacyclopentane fragment influenced by EtMgCl allowed sterically less hindered complex (149), intramolecular transformation of which led to γ-ethyl-substituted α-olefin (150) (Scheme 88).

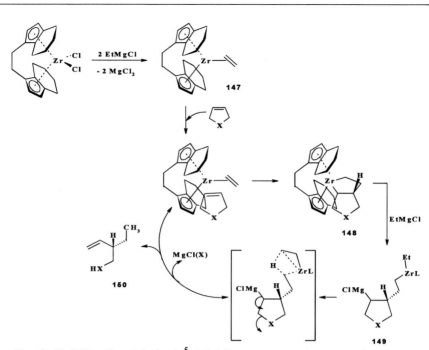

X = O, N C$_9$H$_{19}$; L = 1,2-бис(η^5-4,5,6,7-тетрагидро-1-инденил)этилен

Scheme 88.

Systematic studies of the ethyl- and cyclomagnesiation reaction of olefins with non-activated double bond catalyzed by Cp$_2$ZrCl$_2$, allowed the authors of the work [31, 166, 200] to propose the most probable reaction mechanistic scheme, which clarified the reason of the reaction chemoselectivity in dependence upon the temperature, the selected ethereal solvent and the ratio of initial reagents. According to the proposed scheme primarily the formation of bis-(cyclopentadienyl) zirconium diethyl takes place. The latter transforms to biscyclopentadienylzirconacyclopropane as a result of intramolecular β-elimination. The subsequent insertion of olefin molecule into Zr−C bond provides zirconacyclopentane (145). Oxidative addition of Et$_2$Mg molecule to (145) with subsequent cleavage of Zr−C bond in the latter leads to binuclear Zr−Mg complex (146), intramolecular transformation of which results in the target magnesacyclopentane (103) and regeneration catalytically active Zr complex. Each of these stages of the reaction has been carried out in a stoichiometric version [167].

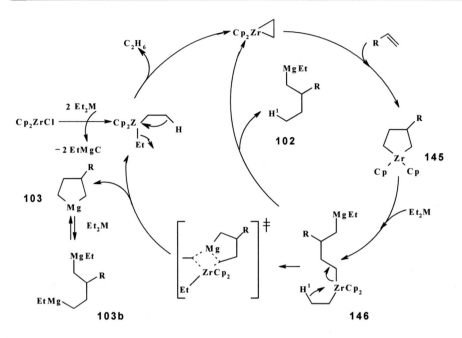

Scheme 89.

Noticeably, the mechanism of the Cp_2ZrCl_2 catalyzed ethylmagnesiation and cyclomagnesiation reactions of olefins with non-activated double bond has been discussed by several groups of investigators almost simultaneously [31, 163, 167–169].

On the example of the reaction of nonconjuagated α,ω-dienes with n-BuMgX or n-Bu$_2$Mg affected by Cp_2ZrCl_2 catalyst (1-10 mole%) leading in ethereal solvents to 1,2-bis(halomagnesiummethyl)carbocycles (151) or 1-(halomagnesiummethyl)-2 methylcarbocycles (152) [169, 173–175] through zirconacyclopentane intermediates (153) [201, 202] it was shown that the yields and the rate of the formation of target products, selectivity and stereochemistry of these reactions depend upon the structure of α,ω-dienes (hepta-1,6-diene, octa-1,7-diene, 9,9-diallylfluorene) and organomagnesium reagent (n-BuMgX, n-Bu$_2$Mg), reaction temperature (0–120 °C), nature of the solvent (Et$_2$O, THF, i-Pr$_2$O, n-Bu$_2$O), and ratio of initial reagents (Scheme 90). The reaction predominantly led to 1,2-bis(n-butylmagnesiummethyl)cycloalkanes (151) while involving n-Bu$_2$Mg in Et$_2$O at 20 °C. The reaction of α,ω-diolefins with n-BuMgX in THF led to the predominant formation of 1-(halomagnesiummethyl)-2-methylcarbocycles (152). Combined yield of the OMCs (151) and (152) in Et$_2$O is more than in THF. The ratio of *trans/cis* stereoisomers of the products (151) and

(152) increased with increasing reaction temperature. Transmetalation of bimetallic complexes (154) by original OMC in catalytic cycle is the key step and determines chemoselectivity in the reaction [174, 175].

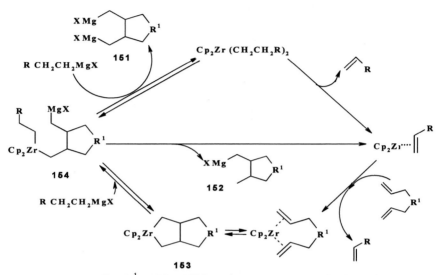

$R = Et$; $R^1 = CH_2$, $(CH_2)_2$, флуоренил; $X = Cl$, Br, n-Bu

Scheme 90.

In order to clarify the mechanism of the above transformations E. Negishi and coworkers have studied the interaction between n-BuMgCl or n-Bu$_2$Mg and 3-bis(η^5-cyclopentadienyl)zirconabicyclo[3.3.0]octane (155) as the probable intermediate in the catalyzed by Cp$_2$ZrCl$_2$ reaction of hepta-1,5-diene with Bu^{n}MgX ($X = Cl$, n-Bu) in different solvents (Scheme 91) [203]. The authors concluded that the reaction of (155) with n-BuMgCl in THF afforded monomagnesium derivative (156). But while using n-BuMgCl (Et$_2$O) and n-Bu$_2$Mg (Et$_2$O or THF) the reaction gave rise to 1,4-dimagnesium derivative (160) as the major product.

This conclusion is based on the fact that Grignard reagents RMgX have monomeric structure in tetrahydrofuran, but dimeric in diethyl ether [204]. The use of dimeric RMgX in Et$_2$O or R$_2$Mg initiates in complexes (158) and (159) intramolecular transfer of alkyl group R from magnesium to zirconium to provide 1,4-dimagnesium compound (160). Similar intramolecular transfer is not feasible when using the monomeric RMgX in THF. Intramolecular transfer of alkyl group R from the magnesium atom in R$_2$Mg to the zirconium atom in complexes (158) and (**159**) occurs at a ratio of (**155**): R_2Mg = 1:1 or 1:2. Monomagnesium unsaturated compounds (**157**) are formed in ~ 10% yield (according to the iodinolysis data) only by the interaction between with n-BuMgCl and hepta-1,6-diene in Et_2O or THF in the presence of catalytic amounts of Cp_2ZrCl_2.

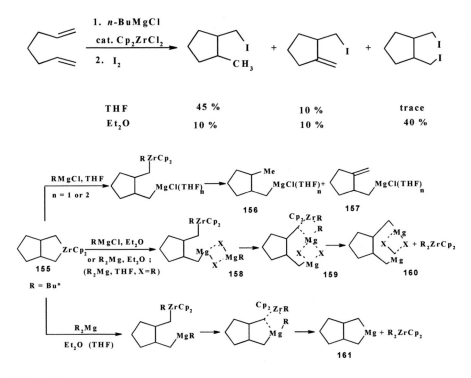

Scheme 91.

Thus, in accordance with the currently available information on the mechanistic routs unsaturated compounds undergo the carbomagnesiation and cyclomagnesiation reactions through the generation of zirconacyclopentane and bimetallic Zr,Mg-complexes as the key intermediates, transformations of which in dependence upon the chosen conditions define the chemoselectivity of the process.

The greatest success, especially in recent years, has been achieved in the decoding mechanism of the catalytic olefin cycloalumination reaction mediated by trialkylalanes and Zr complexes (Scheme 92).

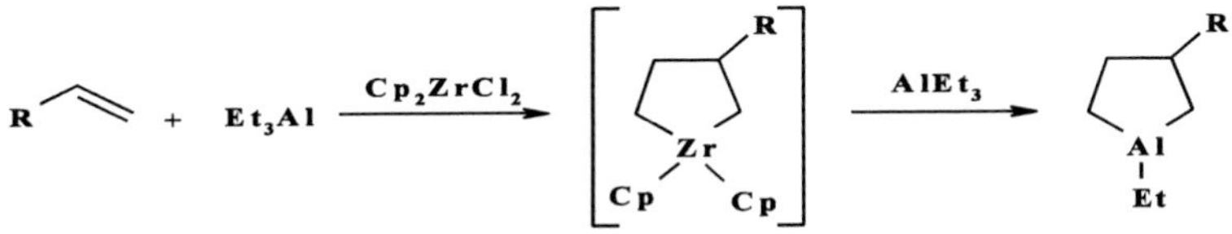

Scheme 92.

Demonstrated in the work [39] supposed simplified mechanistic scheme for the formation of aluminacyclopentanes (13) further has been scrutinized and essentially specified. In particular, mechanism of the cycloalumination reaction of α-olefins with Et_3Al in the presence of Cp_2ZrCl_2 to give aluminacyclopentanes has been tested by means of dynamic ^1H and ^{13}C NMR spectroscopy and simultaneous identification of intermediate Zr–Al bimetallic complexes responsible for the formation of the target aluminacyclopentanes [205]. Furthermore, the authors of the work [206] have succeeded in measuring and quantum-chemical calculation of the formation rate constant of the intermediate complexes and target metallocarbocycles. Mathematical treatment of experimental kinetic data obtained maid it possible to elucidate the stages of regeneration of active catalytic centre and to propose generalized kinetic model for the reaction [207]. The results obtained allowed to conclude that cycloalumination of α-olefins with Et_3Al in the presence of Cp_2ZrCl_2 catalyst proceeds through intermediate bimetallic complex (162). The latter converted into bimetallic bridged complex (163) as a result of ligand exchange between Zr and Al atoms, β-hydride transfer and simultaneous elimination of ethane molecule. Subsequent transformation of unstable complex (164) through the insertion of the starting α-olefin molecule into the Zr–C bond of complex (163) is favorable to the formation of aluminacyclopentane (13) (Scheme 93).

When inquiring about the reaction of trialkylalanes ($AlMe_3$ or $AlEt_3$) with olefins catalyzed by L_2ZrCl_2 [L= Cp, Cp′ (Cp′- η^5-$C_5H_5CH_3$), Cp* (Cp*- η^5-$C_5H_5(CH_3)_5$), Ind (indenyl), Flu (fluorenyl)] the authors of the work [208] have studied the influence of various factors (structure of OAC, π-ligand environment around Zr, nature of solvent, reagent ratio) on the chemoselectivity of the process (Scheme 94). They showed that in the case of $AlMe_3$ the hydrometalation and carbometalation products as well as the products of initial alkene dimerization were formed. In the case of $AlEt_3$ together with the products aforesaid aluminacyclopentanes were detected in the reaction mixture. The all aforesaid allow to propose the generalized mechanism of the interaction between AlR_3 and olefins in the presence of Zr complexes that explained the formation of cyclic and acyclic OAC [208].

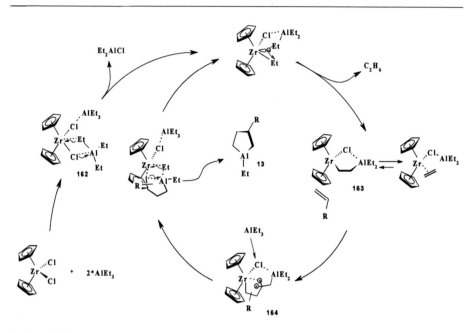

Scheme 93.

The mechanism of the reaction of disubstituted acetylenes with an excess of Et$_3$Al catalyzed by Cp$_2$ZrCl$_2$ has been investigated by Ei-ichi Negishi and coworkers [96]. According to the proposed scheme the reaction proceeds through the formation of bimetallic complex (165). The latter carbometalates the molecule of disubstituted acetylene to give the intermediate (166), which is transformed to the target product (76) and zirconocene ethyl chloride (167). Resultant EtZrCp$_2$Cl (167) in the presence of Et$_3$Al gives complex (168), which is converted to bimetallic complex (165) as a result of β-hydrogen transfer and ethane molecule elimination (Scheme 95).

Mechanism proposed by E. Negishi [96] explains the sequence of the main formation stages for aluminacyclopent-2-enes (**76**) from disubstituted acetylenes, but this mechanism in a less degree correlates with the experimental data obtained by the same authors during the investigations of the interaction between enynes and Et$_3$Al in the presence of Cp$_2$ZrCl$_2$ (Scheme below) [174]. In this case the acetylenic and terminal double bonds should be incorporated into the reaction affording aluminacyclopentenes (79). This fact assumes the formation of the intermediate ethylenezirconocene complex of (165A) type [175, 201−204]. It seems unlikely to obtain (169) through the bimetallic complex (165) with the bridge bond Zr–CH$_2$CH$_2$–Al.

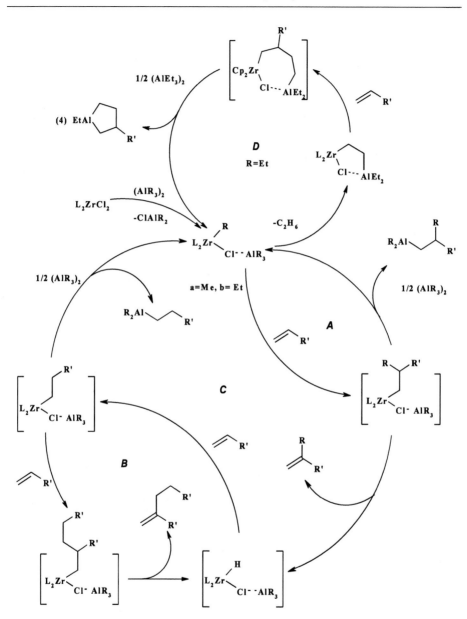

Scheme 94.

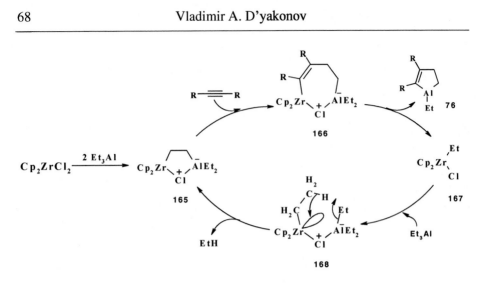

Scheme 95.

In our opinion, the primary formation of (165A) with attached Et$_2$AlCl molecule through the bridge bond Zr---Cl---Al, which can be easily transformed to the target bicyclic aluminacyclopentanes (79), is the most probable rout to synthesize the latter from the complex (168).

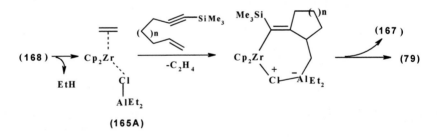

Scheme 96.

The investigations aimed at exploring the mechanisms and defining key reaction intermediates described above, are very important and relevant because they may answer to the question how cyclic OMCs and OACs are formed from acyclic organomagnesium and organoaluminum compounds, and also provide the way to decode general mechanism of action of Ziegler-Natta systems.

Chapter 6

SYNTHESIS OF MACRO METALLACARBOCYCLES THROUGH THE CATALYTIC CYCLOMETALATION REACTION OF UNSATURATED COMPOUNDS

The synthesis of macrocyclic compounds from the simplest olefins, acetylenes and allenes is one of the most promising applications of the catalytic cyclometalation reaction using Mg or Al alkyl derivatives and Ti or Zr complexes as catalysts.

The authors of the work [170, 190] have exploited the regio- and stereoselective cyclometalation reaction, as described in paragraph 4.3, using EtMgBr and the Cp_2TiCl_2 catalyst in the presence of chemically activated Mg [209] for the construction of gigantic metallacarbocycles. The key step of this reaction is the formation of titanacyclopentane intermediates, transmetalation of which with Gringard reagents leads to the appropriate 2,5-dialkylidenemagnesacyclopentanes according to the following Scheme 97:

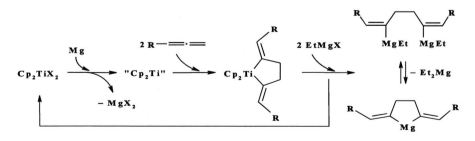

Scheme 97.

Based on the results above the authors of the work [209] supposed that cyclomagnesiation of α,ω-diallenes with Grignard reagents could allow stereoselective intermolecular combined cyclomagnesiation of two or more allene molecules giving rise to macrocyclic organometallic compounds. Building from magnesacyclopentane fragments the latters, after hydrolysis, could be converted into unsaturated hydrocarbon macrocycles, as follows (Scheme 98):

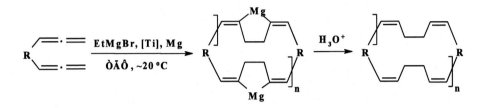

Scheme 98.

Further, the advanced idea has been successfully implemented. Thus, the Mg-containing macrocycle (169)181 with alternating *cis*-double bonds has been synthesized from 1,2,10,11-dodecatetraene and EtMgBr in the presence of chemically activated Mg and Cp_2TiCl_2 catalyst (10 mole %) under mild conditions (r.t., THF, 12 h) in high yield (more than 90%) [209]. Hydrolysis of the resultant (169)181 led to unsaturated macrocycle (170) with Z-double bonds [210]. Hydrogenation or cyclopropanation of the latter allowed to synthesize from α,ω-diallenes through a one pot procedure the corresponding macrocycles (171) and (172) (Scheme 99). As shown, other aliphatic α,ω-diallenes similarly reacted with EtMgBr under above conditions to afford metallacarbocycles.

The elaborated approach allows the authors to obtain the gigantic metallacarbocycles and carbocycles with a number of carbon atoms in the chain from 45 to 60.

These achievements stimulated investigations on intra- and intermolecular cycloalumination of α,ω-diolefines with $RAlCl_2$ in the presence of Cp_2ZrCl_2 to produce carbocycles with annulated aluminacyclopentane fragments [211]. Thus, the interaction between hexa-1,5-diene and $EtAlCl_2$ (1:2 ratio) in the presence of Cp_2ZrCl_2 catalyst (10 mole %) and metallic Mg as an acceptor of halogenide ions in THF (~20 °C, 24 h) was found to provide carbocycles (173), (174) and (175) at a ratio of ~ 4:3:2 in 75 % common yield. Under indicated reaction conditions cycloalumination of octa-1,7-diene with $EtAlCl_2$ led to cyclic OAC (176), (177) and (178) at a ratio of ~ 6:3:1 in 70% total yield. In contrast with hexa-1,5-diene and octa-1,7-diene the cycloalumination reaction of deca-1,9-diene and dodeca-

1,11-diene gave rise to *trans*-3,4-disubstituted aluminacyclopentanes (**179**) in 40–45 % yield (Scheme 100).

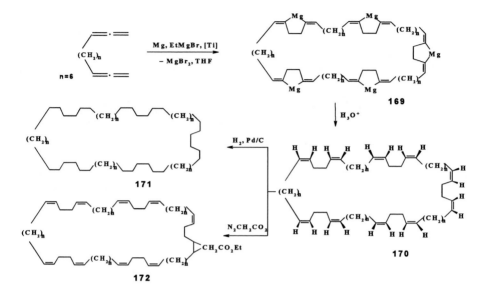

Scheme 99.

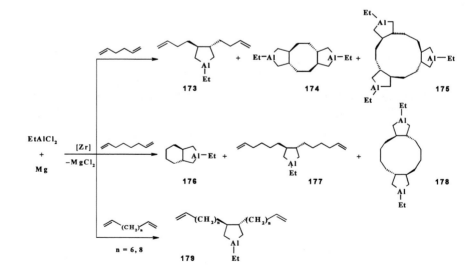

Scheme 100.

Another approach to the synthesis of functionally substituted macrocarbocycles $C_{20}-C_{28}$ is based on the catalytic cyclometalation reactions of cyclic acetylene [212]. For example, the said reaction of cyclooctyne (or cyclododecyne) with RMgX, R_2Mg, $RAlX_2$ or AlX_3 (where X = Cl, Br) in the presence of catalytic amounts of Zr complexes (1–5 mole %) under reaction conditions (ethereal solvent, 20 °C, 3–5 h) was found to provide symmetric tricyclic annulated organomagnesium and organoaluminum compounds. The latters without preliminary isolation were introduced in sequential the cross-coupling reaction with allyl chloride, the cyclometalation reaction with Grignard reagents RMgX or $AlCl_3$, and then in oxidative cleavage reaction that finally led to gigantic carbocyclic ketones in high yields (80–85 %) (Scheme 101).

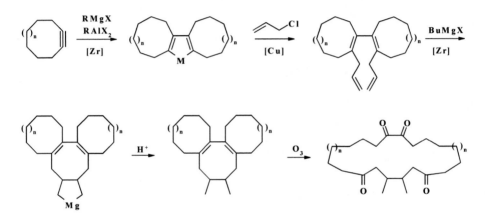

Scheme 101.

Developed by the authors strategy for a synthesis of macrocycles based on the catalytic cyclometalation reaction has a great synthetic potential and opens new perspectives in synthesis of functionally substituted macrocycles of various structure.

The composition and structure of macrocycles have been identified by means of IR, UV, ^1H and ^{13}C NMR spectroscopic methods.

In continuation of research aimed to develop the effective methods for a synthesis of macrocyclic hydrocarbons the authors of the work [213] have elaborated one more way for designing macroheterocycles of a given structure via the cyclomagnesiation reaction of cycloalkynes mediated by Gringard reagents in the presence of a Cp_2ZrCl_2 catalyst to appropriate magnesacyclopentadienes. Cross-coupling of the latters with dihalogen hydrocarbons and subsequent oxidative cleavage of the double bonds led to macrocyclic C_{20}-C_{28} polyketones (Scheme 102).

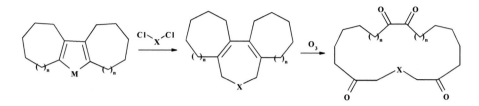

Scheme 102.

So, the surveyed works show the possibility of the catalytic cyclometalation reaction (*Dzhemilev reaction*) exploitation for the construction of the unique **gigantic polyfunctional macrocycles from α,ω-diolefins, α,ω-diallenes and cyclic** acetylenes using available organoaluminum and organomagnesium reagents and transition metal based complexes.

These studies open up fundamentally new opportunities for obtaining selective macrocyclic ionophores, complexing agents for separation, purification and isolation of rare and noble metals, prolongators of medicine, materials for molecular electronics, light-sensitive compositions as well as unique monomers for the synthesis of oligo- and polymeric macromolecules promising for many different areas of industry.

CONCLUSION

The literature data collection presented in this manuscript reflect the most important achievements over the past 15–20 years towards realizing the organoaluminum and organomagnesium synthesis associated with the discovery of the fundamental regioselective 1,2-ethylmagnesiation reaction of olefins with non-activated double bond as well as the catalytic cycloalumination reaction of olefins, acetylenes and 1,2-dienes mediated by Al and Mg alkyls or halogen alkyls in the presence of Ti, Zr or Co catalysts.

The application's inquires around these unique reactions have led to the establishment of the synthetic pathways to the new classes of highly reactive organometallic compounds, namely - aluminacyclopropanes, aluminacyclopropenes, aluminacyclopentanes, aluminacyclopentenes, aluminacyclopenta-2,4-dienes, magnesacyclopentanes, magnesacyclopentenes, magnesacyclopenta-2,4-dienes as well as macrocyclic organoaluminum and organomagnesium compounds.

The elaborated reactions have great synthetic potential as the synthesized Al and Mg metallacycles without preliminary isolation *in situ* can be selectively converted into substituted cyclopropanes, cyclobutanes, thiophanes, thiophenes, phospholanes, cyclopentanols, butan-1,4-diols, gigantic carbocycles in high yields.

Accessibility of initial reagents, simplicity of implementation of transformations, the possibility of application to various olefins, acetylenes and 1,2-dienes make the said reactions (*Dzhemilev reaction*) extremely promising preparative methods for organic and organometallic synthesis.

Without exaggeration, it is possible to tell that the discovered by Professor Dzhemilev catalytic ethylmagnesiation, cycloalumination and also cyclomagnesiation reactions as well as its further elaboration open the new page of organometallic chemistry and provide real prerequisites for designing principally novel chemical technologies.

BIBLIOGRAPHY

[1] Hubel, W.; Braye, E. H. *J. Inorg. Nucl. Chem.* 1959, *10*, 250−268.

[2] Collman, J. P.; Kang, J. W.; Little, W. F.; Sullivan, M. F. *Inorg. Chem.* 1968, *7*, **1298**−1303.

[3] Sears, C. T.; Stone, F. G. A. *J. Organometal. Chem.* 1968, *11*, **644**−646.

[4] Blom, B.; Ceayton, H.; Kilkenni M.; Moss, J. *Advances in Organometallic Chemistry* 2006, *54*, 149−205.

[5] Braye, E. H.; Hubel W. *Chem. and Ind.* 1959, **1250**−1256.

[6] Hubel, W.E.; Braye, H. *US Patent* 3,280,017/1966 (*Chem. Abstr.* 1967, *66*, 2463d).

[7] Muller, E.; Beissner, C.; Jakle, H.; Langer, E.; Muhm, H.; Odenigbo, G.; Sauerbier, M.; Segnitz, A.; Streichfuss, D.; Tomas, R. *Annalen.*, 1972, *754*, **64**−89.

[8] Watt, G. W.; Dummond, F. O. *J. Am. Chem. Soc.* 1970, *92*, **826**−828.

[9] Seetz, J. W. F. L.; Hartog, F. A.; Böhm, H. P.; Blomberg, C.; Akkerman, O. S.; Bickelhaupt, F. *Tetrahedron Lett.* 1982, *23*, **1497**−1500.

[10] Denise, B.; Fauvargue, J.-F.; Ducom, J. *Tetrahedron Lett.* 1970, *11*, **335**−338.

[11] Holtkamp, H.C.; Schat, G.; Blomberg, C.; Bickelhaupt, F. *J. Organomet. Chem.* 1982, *240*, 1−8.

[12] Freijee, F. J. M.; Schat, G.; Akkerman, O.S.; Bickelhaupt, F. *J. Organomet. Chem.* 1982, *240*, 217−227.

[13] Freijee, F. J. M.; Van der Wal, G.; Schat, G.; Akkerman, O.S.; Bickelhaupt, F. *J. Organomet. Chem.* 1982, *240*, 229−238.

[14] Fujita, K.; Ohnuma, Y.; Yasuda, H.; Tani, H. *J. Organomet. Chem.* 1976, *113*, 201−213.

[15] Wreford, S. S.; Whitney, J. F. *Inorg. Chem.* 1981, *20*, **3918**−3924.

[16] Xiong H.; Rieke, R. D. *J. Org. Chem.* 1989, *54*, **3247**−3249.

[17] Moiseenkov, A. M.; Czeskis, B. A.; Semenovsky, A. V. *Tetrahedron Lett.* 1980, *21*, 853–856.

[18] Dzhemilev, U. M.; Ibragimov, A. G.; Vostrikova, O. S.; Tolstikov, G. A. *Izv. Akad. Nauk SSSR, Ser. Khim.* 1979, 2071–2074 [*Bull. Akad. Sci. USSR, Div. Chem. Sci.* 1979, *28* (Engl. Transl.)].

[19] Dzhemilev, U. M.; Ibragimov, A. G.; Tolstikov, G. A.; Vostrikova, O. S.; Zelenova, L. M. *Zhurn. Organ. Khimii* 1981, *17*, 2313–2319 (in Russian).

[20] Vostrikova, O. S.; Ibragimov, A. G.; Tolstikov, G. A.; Zelenova, L. M.; Dzhemilev, U. M. *Izv. Akad. Nauk SSSR, Ser. Khim.* 1981, 1410–1412 [*Bull. Akad. Sci. USSR, Div. Chem. Sci.* 1981, *30*, 1132 (Engl. Transl.)].

[21] Vostrikova, O. S.; Ibragimov, A. G.; Tolstikov, G. A.; Zelenova, L. M.; Dzhemilev, U. M. *Izv. Akad. Nauk SSSR, Ser. Khim.* 1980, 2320–2322 [*Bull. Akad. Sci. USSR, Div. Chem. Sci.* 1980, *29*, 1638 (Engl. Transl.)].

[22] Dzhemilev, U. M.; Vostrikova, O. S.; Tolstikov, G. A.; Ibragimov, A. G. *Izv. Akad. Nauk SSSR, Ser. Khim.* 1979, 2626–2627 [*Bull. Akad. Sci. USSR, Div. Chem. Sci.* 1979, *28*, 2441 (Engl. Transl.)].

[23] Dzhemilev, U. M.; Ibragimov, A. G.; Vostrikova, O. S.; Tolstikov, G. A.; Zelenova, L. M. *Izv. Akad. Nauk SSSR, Ser. Khim.* 1981, 361–364 [*Bull. Akad. Sci. USSR, Div. Chem. Sci.* 1981, *30*, 281 (Engl. Transl.)].

[24] Dzhemilev, U. M.; Ibragimov, A. G.; Vostrikova, O. S.; Tolstikov, G. A. *Izv. Akad. Nauk SSSR, Ser. Khim.* 1985, 207–209 [*Bull. Akad. Sci. USSR, Div. Chem. Sci.* 1985, 34, 196 (Engl. Transl.)].

[25] Bradford P. Mundy, Michael G. Ellerd, Frank G. Favaloro, *Name Reactions and Reagents in Organic Synthesis*, Second Edition, John WileyandSons, 2005, P. 286.

[26] Negishi, E. *Bull. Chem. Soc. Jap.*, 2007, *80*, 233–257.

[27] Quntar, Abed Al Aziz; Srebnik, M. *J. Org. Chem.*, 2006, *71*, 730–733.

[28] Fang, H.; Zhao, C.; Li, G.; Xi, Z. *Tetrahedron*, 2003, *59*, 3779–3786.

[29] Dzhemilev, U. M.; Vostrikova, O. S.; Sultanov, R. M. *Izv. Akad. Nauk SSSR, Ser. Khim.* 1983, 218–220 [*Bull. Akad. Sci. USSR, Div. Chem. Sci.* 1983, *32*, 193 (Engl. Transl.)].

[30] Dzhemilev, U. M.; Vostrikova, O. S. *J. Organomet. Chem.* 1985, *285*, 43–51.

[31] Lewis, D. P.; Muller, P. M.; Whitby, R. J.; Jones, R. V. *Tetrahedron Lett.* 1991, *32*, 6797–6800.

[32] Dzhemilev, U. M.; Ibragimov, A. G.; Zolotarev, A. P.; Muslukhov, R. R.; Tolstikov, G. A. *Izv. AN SSSR. Ser. Khim.* 1990, 2831–2841 [*Bull. Akad. Sci. USSR, Div. Chem. Sci.* 1990, *39*, 2570 (Engl. Transl.)].

[33] Dzhemilev, U. M.; Vostrikova, O. S.; Tolstikov, G. A. *Uspekhi khimii*, 1990, *59*, 1972−2002.

[34] Vostrikova, O. S.; Ibragimov, A. G.; Sultanov, R. M.; Dzhemilev, U. M. *Metalloorg. Khim.* 1992, *5*, 782−793 [*Organomet. Chem. USSR* 1992, *5* (Engl. Transl.)].

[35] Cristian, A.-M. C.; Krylov, A. I. *J. Chem. Phys.*, 2003, *118*, 10912−10918. [Dewar, M. J. S. *Bull. Soc. Chim. Fr.*, 1951, 18; Chatt J. and Dunkanson, L. A. *J. Chem. Soc.* 1953, 2939].

[36] Dzhemilev, U. M.; Ibragimov, A. G.; Morozov, A. B. *Mendeleev Commun.* 1992, 26−28.

[37] Dzhemilev U. M.; Ibragimov, A. G. *J. Organomet. Chem.* 1994, *466*, 1−4.

[38] Dzhemilev, U. M. *Tetrahedron* 1995, *51*, 4333−4346.

[39] Dzhemilev, U. M.; Ibragimov, A. G.; Zolotarev, A. P.; Muslukhov, R. R.; Tolstikov, G. A. *Izv. Akad. Nauk SSSR, Ser. Khim.* 1989, 207−208 [*Bull. Acad. Sci. USSR, Div. Chem. Sci.* 1989, *38*, 194 (Engl. Transl.)].

[40] Dzhemilev U. M.; Ibragimov, A. G. *Izv. Akad. Nauk SSSR, Ser. Khim.* 1998, 816−823 [*Russ. Chem. Bull.* 1998, *47*, 786 (Engl. Transl.)].

[41] Dzhemilev, U. M.; Ibragimov, A. G.; Khafizova, L. O.; Rusakov, S. V.; Khalilov, L. M. *Mendeleev Commun.* 1997, 198−199.

[42] Ibragimov, A. G.; Khafizova, L. O.; Gil'fanova, G. N.; Dzhemilev, U. M. *Izv. Akad. Nauk, Ser. Khim.* 2002, 2095−2100 [*Russ.Chem. Bull., Int. Ed.* 2002, *51*, 2255 (Engl. Transl.)].

[43] Dzhemilev, U. M.; Ibragimov, A. G.; Ramazanov, I. R.; Luk'yanova, M. P.; Sharipova, A. Z. *Izv. Akad. Nauk, Ser. Khim.* 2001, 465−468 [*Russ. Chem. Bull., Int. Ed.* 2001, *50*, 484 (Engl. Transl.)].

[44] Dzhemilev, U. M., Ibragimov, A. G.; Khafizova, L. O.; Yakupova, L. R.; Khalilov, L. M. *Zhurn. Organ. Khimii* 2005, *41*, 684−689 [*Russ. J. Org. Chem.* 2005, 41, 667 (Engl. Transl.)].

[45] Dzhemilev, U. M.; Ibragimov, A. G.; Ramazanov, I. R.; Khalilov, L. M. *Izv. Akad. Nauk, Ser. Khim.* 1997, 2269−2270 [*Russ. Chem. Bull.* 1997, *46*, 2150 (Engl. Transl.)].

[46] Horn, D. E. V.; Negishi, E. *J. Am. Chem. Soc.* 1978, *100*, 2252−2254.

[47] Vara Prosad, J. V. N.; Pillai, C. N. *J. Organomet. Chem.* 1983, *259*, 1−30.

[48] Normant, J. F.; Alexakis, A. *Synthesis* 1981, 841−870.

[49] Negishi, E. *Acc. Chem. Res.* 1987, *20*, 65−72.

[50] Negishi, E.; Kondakov, D.Y. *Chem. Soc. Rev.* 1996, 417−426.

[51] *Comprehensive Organic Transformations. A Guide to Functional Group Preparations.* (Ed. by R.C. Larock). VCH, New York, 1999.

[52] Lin, S.-H. *J. Org. Chem.* 1977, *42*, 3209−3210.

[53] Farady, L.; Bencze, L.; Marco, L. *J. Organomet. Chem.* 1967, *10*, 505–510.

[54] Farady, L.; Marco, L. *J. Organomet. Chem.* 1971, *28*, 159–165.

[55] Farady, L.; Bencze, L.; Marko, L. *J. Organomet. Chem.* 1969, *17*, 107–116.

[56] Vostrikova, O. S.; Dzhemilev, U. M.; Sultanov, R. M. *Izv. Akad. Nauk SSSR, Ser. Khim.* 1985, 1430–1433 [*Bull. Akad. Sci. USSR, Div. Chem. Sci.* 1985, *34*, 1449 (Engl. Transl.)].

[57] Dzhemilev, U. M.; Vostrikova, O. S. *J. Organomet. Chem.* 1985, *285*, 43–51.

[58] Hoveyda, A. H.; Xu, Z. *J. Am. Chem. Soc.* 1991, *113*, 5079–5080.

[59] Houri, A. F.; Didiuk, M. T.; Xu, Z.; Horan, N. R.; Hoveyda, A. H. *J. Am. Chem., Soc.* 1993, *115*, 6614–6624.

[60] Hoveyda, A. H.; Xu, Z.; Morken, J. P.; Houri, A. F. *J. Am. Chem. Soc.* 1991, *113*, 8950–8952.

[61] Hoveyda, A. H.; Morken. J. P. *J. Org. Chem.* 1993, *58*, 4237–4244.

[62] Wild, F. R. W. P.; Wasiucionek, M.; Huttner, G.; Brintzinger, H. H. *J. Organomet. Chem.* 1985, *288*, 63–67.

[63] Schafer, A.; Karl, E.; Zsolnai, L.; Huttner, G.; Brintzinger, H. H. *J. Organomet. Chem.* 1987, *328*, 87–99.

[64] Chin, B.; Buchwald, S.L. *J. Org. Chem.* 1997, *62*, 2267–2268.

[65] Bell, L.; Whitby, R. J.; Jones, R. V. H.; Standen, M. C. H. *Tetrahedron Lett.* 1996, *37*, 7139–7142.

[66] Morken, J. P.; Didiuk, M. T.; Hoveyda, A. H. *J. Am. Chem. Soc.* 1993, *115*, 6997–6998.

[67] Morken, J. P.; Didiuk, M. T.; Visser, M. S.; Hoveyda, A. H. *J. Am. Chem. Soc.* 1994, *116*, 3123–3124.

[68] Visser, M. S.; Heron, N. M.; Didiuk, M. T.; Sagal, J. F.; Hoveyda, A. H. *J. Am. Chem. Soc.* 1996, *118*, 4291–4298.

[69] Visser, M. S.; Harrity, J. P. A.; Hoveyda, A. H. *J. Am. Chem. Soc.* 1996, *118*, 3779–3780.

[70] Adams, J. A.; Heron, N. M.; Koss, A.-M.; Hoveyda, A. H. *J. Org. Chem.* 1999, *64*, 854–860.

[71] Johannes, C. W.; Visser, M. S.; Weatherhead, G. S.; Hoveyda, A. H. *J. Am. Chem. Soc.* 1998, *120*, 8340–8347.

[72] Hoveyda, A. H.; Morken, J. P. *Angew. Chem. Int. Ed. Engl.* 1996, *35*, 1262–1284.

[73] Hoveyda, A. H. *In Titanium and Zirconium in Organic Synthesis.* (Ed. By I. Marek). Wiley-VCH: Weinheim 2002, 181.

[74] Hoveyda, A. H.; Morken, J. P. *Angew. Chem.* 1996, *108*, 1378–1386.

[75] Hoveyda, A. H.; Heron, N. M. *In Comprehensive Asymmetric Catalysis* (Eds.: E.N. Jacobsen, A. Pfaltz, H. Yamamoto). Springer, Berlin, 1999, 431.

[76] Visser, M. S.; Hoveyda, A. H. *Tetrahedron* 1995, *51*, 4383–4394.

[77] Bell, L.; Whitby, R. J.; Jones, V. H.; Standen, M. C. H. *Tetrahedron Lett.* 1996, *37*, 7139–7142.

[78] Bell, L.; Brookings, D. C.; Dawson, G. J.; Whitby, R. J. *Tetrahedron* 1998, *54*, 14617–14634.

[79] Marek, I. *J. Chem. Soc., Perkin Trans.1* 1999, 535–539.

[80] Houri, F.; Xu, Z. –M.; Cogan, D. A.; Hoveyda, A. H. *J. Am. Chem. Soc.* 1995, *117*, 2943–2944.

[81] Xu, Z.; Johannes, C.W.; Houri, A.F.; La, D. S.; Cogan, D. A.; Hofinela, G. E.; Hoveyda, A.H. *J. Am. Chem. Soc.* 1997, *119*, 10302–10316.

[82] Xu, Z.; Johannes, C. W.; Salman, S. S.; Hoveyda, A. H. *J. Am. Chem. Soc.* 1996, *118*, 10926–10927.

[83] Hoveyda, A. H.; Morken, J. P.; Houri, A. F.; Xu, Z. *J. Am. Chem. Soc.* 1992, *114*, 6692–6697.

[84] Didiuk, M. T.; Johannes, C. W.; Morken, J. P.; Hoveyda, A.H. *J. Am. Chem. Soc.* 1995, *117*, 7097–7104.

[85] Tebben, G.-D.; Rauch, K.; Stratmann, C.; Williams, C. M.; deMeijere, A. *Organic Letters* 2003, *5*, 483–485.

[86] Gagneur, S.; Montchamp, J. L.; Negishi, E. *Organometallics* 2000, *19*, 2417–2419.

[87] Dzhemilev, U. M.; Ibragimov, A. G. *Usp. Khim.* 2000, *69*, 134–149 [*Russ. Chem. Rev.* 2000, *69*, 121 (Engl. Transl.)].

[88] Ibragimov, A. G.; Khafizova, L. O.; Gil'fanova, G. N.; Yakupova, L. R.; Borisova, A. L.; Dzhemilev, U. M. *Izv. Akad. Nauk, Ser. Khim.* 2003, 2302–2306 [*Russ. Chem. Bull., Int. Ed.* 2003, *52*, 2434 (Engl. Transl.)].

[89] Kondakov, D. Y.; Negishi, E. *J. Am. Chem. Soc.* 1996, *118*, 1577–1578.

[90] Dawson, G.; Durrant, C. A.; Kirk, G. G.; Whitby, R. J.; Jones, R. V. H.; Standen, M. C. H. *Tetrahedron Lett.* 1997, *38*, 2335–2338.

[91] Khalilov, L. M.; Parfenova, L. V.; Pechatkina, S. V.; Ibragimov, A. G.; Genet, J. P.; Dzhemilev, U. M.; Beletskaya, I. P. *J. Organomet. Chem.* 2004, *689*, 444–453.

[92] Ibragimov, A. G.; Zagrebel'naya, I. V.; Satenov, K. G.; Khalilov, L. M.; Dzhemilev, U. M. *Izv. Akad. Nauk, Ser. Khim.* 1998, 712–715 [*Russ. Chem. Bull.* 1998, *47*, 691 (Engl. Transl.)].

[93] Ramazanov, I. R.; D'yachenko, L. I.; Ibragimov, A. G.; Dzhemilev, U. M. *Izv. Akad. Nauk, Ser. Khim.* 2002, 770–772 [*Russ. Chem. Bull., Int. Ed.* 2002, *51*, 833 (Engl. Transl.)].

[94] Millward, D. B.; Cole, A. P.; Waymouth, R. M. *Organometalics* 2000, *19*, 1870−1878.

[95] Ibragimov, A. G.; Khafizova, L. O.; Satenov, K. G.; Khalilov, L. M.; Yakovleva, L. G., Rusakov, S. V.; Dzhemilev, U. M. *Izv. Akad. Nauk, Ser. Khim.* 1999, 1594−1600 [*Russ. Chem. Bull.* 1999, *48*, 1574 (Engl. Transl.)].

[96] Negishi, E.; Kondakov, D. Y.; Choueiry, D.; Kasai, K.; Takahashi, T. *J. Am. Chem. Soc.* 1996, *118*, 9577−9588.

[97] Kaminsky, W.; Sinn, H. *Liebigs Ann. Chem.* 1975, **424**−437.

[98] Kaminsky, W.; Vollmer, H. J. *Liebigs Ann. Chem.* 1975, **438**−448.

[99] Kaminsky, W.; Kopf, J.; Sinn, H.; Vollmer, H. J. *Angew. Chem. Int. Ed. Engl.* 1976, *15*, 629−630.

[100] Dzhemilev, U. M. et al. RF Patent 2157812; *Byull. Izobret.* [*Bulletin of Inventions*], 2000, No.29 (in Russian).

[101] Dzhemilev, U. M.; Ibragimov, A. G.; Zolotarev, A. P.; Khalilov, L. M.; Muslukhov, R. R. *Izv. Akad. Nauk SSSR, Ser. Khim.* 1992, **386**−391 [*Bull. Russ. Akad. Sci., Div. Chem. Sci.* 1992, *41*, 300 (Engl. Transl.)].

[102] Dzhemilev, U. M.; Ibragimov, A. G.; Khafizova, L. O.; Khalilov, L. M.; Vasiliev, Yu. V.; Tuktarov, R. F.; Tomilov, Yu. V.; Nefedov, O. M. *Izv. Akad. Nauk, Ser. Khim.* 1999, 572−574 [*Russ. Chem. Bull.* 1999, *48*, 567 (Engl. Transl.)].

[103] Dzhemilev, U. M.; Ibragimov, A. G.; Morozov, A. B.; Khalilov, L. M.; Muslukhov, R. R.; Tolstikov, G. A. *Izv. Akad. Nauk SSSR, Ser. Khim.* 1991, 1141−1144 [*Bull. Akad. Sci. USSR, Div. Chem. Sci.* 1991, *40*, 1022 (Engl. Transl.)].

[104] Dzhemilev, U. M.; Ibragimov, A. G.; Morozov, A. B.; Muslukhov, R. R.; Tolstikov, G. A. *Izv. Akad. Nauk SSSR, Ser. Khim.* 1991, 1607−1609 [*Bull. Akad. Sci. USSR, Div. Chem. Sci.* 1991, *40*, 1425 (Engl. Transl.)].

[105] Dzhemilev, U. M.; Ibragimov, A. G.; Morozov, A. B.; Muslukhov, R. R.; Tolstikov, G. A. *Izv. Akad. Nauk, Ser. Khim.* 1992, 1393−1397 [*Bull. Akad. Sci. USSR, Div. Chem. Sci.* 1992, *41*, 1089 (Engl. Transl.)].

[106] Ibragimov, A.G.; Ermilova, O.E.; Kunakova, R. V.; Islamgulova, A. Z.; Dzhemilev, U. M. In IXth International Symposium on Organometallic Chemistry (UPAC), *Abstracts of Reports*, Gottingen (Germany) 1997, 430.

[107] Negishi, E.; Takahashi, T. *Synthesis* 1988, *1*, 1−44.

[108] Negishi, E. *Pure Appl. Chem.* 1992, *64*, 323−335.

[109] Negishi, E.; Holmes, S. J.; Tour, J. M.; Miller, J. A. *J. Am. Chem. Soc.* 1985, *107*, **2568**−2569.

[110] Rajan Babu, T. V.; Myent, W. A.; Taber, D. F.; Fagan, P. J. *J. Am. Chem. Soc.* 1988, *110*, **7128**−7135.

[111] Dzhemilev, U. M.; Ibragimov, A. G.; Azhgaliev, M. N.; Muslukhov, R. R. *Izv. Akad. Nauk, Ser. Khim.* 1995, 1561−1567 [*Russ. Chem. Bull.* 1995, *44*, 1501 (Engl. Transl.)].

[112] Ibragimov, A. G.; Khafizova, L. O.; Zagrebel'naya, I. V.; Parfenova, L. V.; Sultanov, R. M.; Khalilov, L. M.; Dzhemilev, U. M. *Izv. Akad. Nauk, Ser. Khim.* 2001, 280−284 [*Russ. Chem. Bull. Int. Ed.* 2001, *50*, 292 (Engl. Transl.)].

[113] Muslukhov, R. R.; Khalilov, L. M.; Zolotarev, A. P.; Morozov, A. B.; Ibragimov, A. G.; Dzhemilev, U. M.; Tolstikov, G. A. *Izv. Akad. Nauk, Ser. Khim.* 1992, 2110−2116 [*Bull. Russ. Akad. Sci., Div. Chem. Sci.* 1992, *41*, 1646 (Engl. Transl.)].

[114] Muslukhov, R. R.; Khalilov, L. M.; Ibragimov, A. G.; Zolotarev, A. P.; Dzhemilev, U. M. *Izv. Akad. Nauk, Ser. Khim.* 1992, 2742−2750 [*Bull. Russ. Akad. Sci., Div. Chem. Sci.* 1992, *41*, 2172 (Engl. Transl.)].

[115] Ibragimov, A. G.; Khafizova, L. O.; Yakovleva, L. G.; Nikitina, E. V.; Satenov, K. G.; Khalilov, L. M.; Dzhemilev, U. M. *Izv. Akad. Nauk, Ser. Khim.* 1999, 778−784 [*Russ. Chem. Bull.* 1999, *48*, 774 (Engl. Transl.)].

[116] Dzhemilev, U. M.; Ibragimov, A. G.; Morozov, A. B.; Tolstikov, G. A. Tezisy dokladov Vsesoyuznoi konferentsii po metalloorganicheskoi khimii, Riga, [In All-Russian Conference on Organometallic Chemistry, Abstracts], 1991, *5*, 101.

[117] Dzhemilev, U. M.; Ibragimov, A. G.; Zolotarev, A. P.; Muslukhov, R. R. *Izv. Akad. Nauk, Ser. Khim.* 1992, 382−385 [*Bull. Russ. Akad. Sci., Div. Chem. Sci.* 1992, *41*, 297 (Engl. Transl.)].

[118] Dzhemilev, U. M.; Ibragimov, A. G.; Zolotarev, A. P. *Mendeleev Commun.* 1992, 28−29.

[119] Dzhemilev, U. M. et al. RF Patent Documents: 2109717; 2109718; *Byull. Izobret.* [*Bulletin of Inventions*], 1998, No.12 (in Russian).

[120] Dzhemilev, U. M.; Ibragimov, A. G.; Zolotarev, A. P.; Muslukhov, R. R.; Tolstikov, G. A. *Izv. Akad. Nauk SSSR, Ser. Khim.* 1990, 1190−1191 [*Bull. Akad. Sci. USSR, Div. Chem. Sci.* 1990, *39*, 1071 (Engl. Transl.)].

[121] Dzhemilev, U. M.; Ibragimov, A. G.; Khafizova, L. O.; Parfenova, L. V.; Yalalova, D. F.; Khalilov, L. M. *Izv. Akad. Nauk, Ser. Khim.* 2001, 1394−1397 [*Russ. Chem. Bull., Int. Ed.* 2001, *50*, 1465 (Engl. Transl.)].

[122] Dzhemilev, U. M.; Ibragimov, A. G.; Khafizova, L. O.; Ramazanov, I. R.; Yalalova, D. F.; Tolstikov, G. A. *J. Organomet. Chem*, 2001, *636*, 76−81.

[123] Dzhemilev, U. M.; Ibragimov, A. G.; Zolotarev, A. P.; Muslukhov, R. R.; Tolstikov, G. A. *Izv. Akad. Nauk SSSR, Ser. Khim.* 1989, 2152 [*Bull. Akad. Sci. USSR, Div. Chem. Sci.* 1989, *38*, 1981 (Engl. Transl.)].

[124] Dzhemilev, U. M.; Ibragimov, A. G.; Azhgaliev, M. N.; Zolotarev, A. P.; Muslukhov, R. R. *Izv. Akad. Nauk, Ser. Khim.* 1994, 273−275 [*Russ. Chem. Bull.* 1994, *43*, 252 (Engl. Transl.).

[125] Ibragimov, A. G.; Zolotarev, A. P.; Muslukhov, R. R.; Lomakina, S. I.; Dzhemilev, U. M. *Izv. Akad. Nauk, Ser. Khim.* 1995, 118−120 [*Russ. Chem. Bull.* 1995, *44*, 113 (Engl. Transl.)].

[126] Dzhemilev, U. M.; Ibragimov, A. G.; Zolotarev, A. P.; Tolstikov, G. A. *Izv. Akad. Nauk SSSR, Ser. Khim.* 1989, 1444 [*Bull. Akad. Sci. USSR, Div. Chem. Sci.* 1989, *38*, 1324 (Engl. Transl.)].

[127] Dzhemilev, U. M., Ibragimov, A. G.; Azhgaliev, M. N.; Muslukhov, R. R. *Izv. Akad. Nauk, Ser. Khim.* 1994, 276−278 [*Russ.Chem. Bull.* 1994, *43*, 255 (Engl. Transl.)].

[128] Dzhemilev, U. M.; Ibragimov, A. G.; Gilyazev, R. R.; Khafizova, L. O. *Tetrahedron* 2004, *60*, 1281−1286.

[129] Dzhemilev, U. M.; Ibragimov, A. G.; Khafizova, L. O.; Gilyazev, R. R.; D'yakonov, V. A. *Izv. Akad. Nauk, Ser. Khim.* 2004, 130−133 [*Russ. Chem. Bull. Int. Ed.* 2004, *53*, 133 (Engl. Transl.)].

[130] Dzhemilev, U. M.; Ibragimov, A. G.; Khafizova, L. O.; Gilyazev, R. R.; Makhamatkhanova, A. L. *Zhurn. Organ. Khimii* 2007, *43*, 352−356 [*Russ. J. Org. Chem.* 2007, 43, 347 (Engl. Transl.)].

[131] Khafizova, L. O.; Ibragimov, A. G.; Yalalova, D. F.; Borisova, A. L.; Khalilov, L. M.; Dzhemilev, U. M. *Izv. Akad. Nauk, Ser. Khim.* 2003, 1905−1909 [*Russ.Chem. Bull., Int. Ed.* 2003, *52*, 2012 (Engl. Transl.)].

[132] Odinokov, V. N.; Ishmuratov, G. Yu.; Kharisov, R. Ya.; Ibragimov, A. G.; Sultanov, R. M.; Dzhemilev, U. M.; Tolstikov, G. A. *Khim. Prir. Soedin.* 1989, 272−275 [*Chem. Nat. Compd.* 1989 (Engl. Transl.)].

[133] Odinokov, V. N.; Ishmuratov, G. Yu.; Ibragimov, A. G.; Yakovleva, M. P.; Zolotarev, A. P.; Dzhemilev, U. M.; Tolstikov, G. A. *Khim. Prir. Soedin.* 1992, 567−571 [*Chem. Nat. Compd.* 1992 (Engl. Transl.)].

[134] Dzhemilev, U. M.; Ibragimov, A. G.; Kunakova, R. V.; Minsker, D. L.; Yusupov, Z. A. In XIIth Fechem Conference on Organometallic Chemistry, Abstracts, Prague 1997, 52.

[135] Ibragimov, A. G.; Ermilova, O. E.; Kunakova, R. V.; Islamgulova, A. Z.; Dzhemilev, U. M. In XIIth Fechem Conference on Organometallic Chemistry, Abstracts, Prague 1997, 47.

[136] D'yakonov, V. A.; Finkelshtein, E. Sh.; Ibragimov, A. G. *Tetrahedron Lett.* 2007, *48*, 8583−8586.

[137] D'yakonov, V. A.; Trapeznikova, O. A.; Ibragimov, A. G.; Dzhemilev, U. M. *Izv. Akad. Nauk, Ser. Khim.* 2009, 926−932 [*Russ. Chem. Bull., Int. Ed.* 2009 (Engl. Transl.)].

[138] D'yakonov, V. A.; Ibragimov, A. G.; Khalilov, L. M.; Makarov, A. A.; Timerkhanov, R. K.; Tuktarova, R. A.; Trapeznikova, O. A.; Galimova, L. F. *Khimiya Geterotsiklicheskikh Soedinenii* 2009, *501*, 393−402 [*Chem. Heterocyclic Comp.* 2009, *45*, 395].

[139] Ibragimov, A. G.; Khafizova, L. O.; Khalilov, L. M.; Dzhemilev, U. M. In IXth Symposium on Organometallic Chemistry (UPAC), Abstracts of Reports, Gottingen (Germany) 1997, 428.

[140] Shur, V. B.; Berkovich, E. G.; Vol'pin, M. E.; Lorenz, B.; Wahren, M. *J. Organomet. Chem.* 1982, *228*, C36−C38.

[141] Dzhemilev, U. M.; Ibragimov, A. G.; Khafizova, L. O.; Khalilov, L. M.; Vasiliev, Yu. V.; Tomilov, Yu. V. *Izv. Akad. Nauk, Ser. Khim.* 2001, 285−288 [*Russ.Chem. Bull., Int. Ed.* 2001, *50*, 297 (Engl. Transl.)].

[142] Burlakov, V. V.; Usatov, A. V.; Lyssenko, K. A.; Antipin, M. Yu.; Novikov, Yu. N.; Shur, V. B. *Eur. J. Inorg. Chem.* 1999, 1855−1857.

[143] Khafizova, L. O.; Ibragimov, A. G.; Gil'phanova, G. N.; Khalilov, L. M.; Dzhemilev, U. M. *Izv. Akad. Nauk, Ser. Khim.* 2001, 2089−2093 [*Russ.Chem. Bull., Int. Ed.* 2001, *50*, 2188 (Engl. Transl.)].

[144] D'yakonov, V. A.; Zinnurova, R. A.; Timerkhanov, R. K.; Ibragimov, A. G.; Dzhemilev, U. M. In Eighth Tetrahedron Symposium "Challenges in Organic Chemistry", Abstracts, Berlin 2007, P1.81.

[145] D'yakonov, V. A.; Timerkhanov, R. K.; Ibragimov, A. G.; Dzhemilev, U. M. *Izv. Akad. Nauk, Ser. Khim.* 2007, 2156−2159 [*Russ.Chem. Bull., Int. Ed.* 2007, *56*, 2232 (Engl. Transl.)].

[146] D'yakonov, V. A.; Timerkhanov, R. K.; Ibragimov, A. G.; Tumkina, T. V.; Popod'ko, N. R.; Dzhemilev, U. M. *Tetrahedron Lett.* 2009, *50*, 1270−1272.

[147] D'yakonov, V. A.; Tumkina, T. V.; Dzhemilev, U. M. *Izv. Akad. Nauk, Ser. Khim.* 2009 (in press) [*Russ.Chem. Bull., Int. Ed.* 2009 (Engl. Transl.)].

[148] In *Vazhneishie Resul'taty v Oblasti Estestvennykh i Obshchestvennykh Nauk za 1990 god* (Otchet Akad. Nauk SSSR), Moskva, 1991 [*Major Results in the Field of Natural and Social Sciences for 1990* (Report of Academy of Sciences of the USSR), Moscow, 1991] 44−48.

[149] Dzhemilev, U. M.; Ibragimov, A. G.; Zolotarev, A. P. *Mendeleev Commun.* 1992, 135−136.

[150] Muslukhov, R. R.; Khalilov, L. M.; Ramazanov, I. R.; Sharipova, A. Z.; Ibragimov, A. G.; Dzhemilev, U. M. *Izv. Akad. Nauk, Ser. Khim.* 1997, 2194–2197 [*Russ. Chem. Bull.* 1997, *46*, 2082 (Engl. Transl.)].

[151] Dzhemilev, U. M.; Ibragimov, A. G.; Ramazanov, I. R.; Luk'yanova, M. P.; Sharipova, A. Z.; Khalilov, L. M. *Izv. Akad. Nauk, Ser. Khim.* 2000, 1092–1095 [*Russ. Chem. Bull.* 2000, *49*, 1086 (Engl. Transl.).

[152] Negishi, E.; Montchamp, J.-L.; Anastasia, L.; Elizarov, A.; Choueiry, D. *Tetrahedron Lett.* 1998, *39*, 2503–2506.

[153] Ramazanov, I. R.; Ibragimov, A. G.; Dzhemilev, U. M. *Zhurn. Organ. Khimii* 2008, *44*, 793–796 [*Russ. J. Org. Chem.* 2008, *44*, 781 (Engl. Transl.)].

[154] Khafizova, L. O.; Yakupova, L. R.; Ibragimov, A. G.; Dzhemilev, U. M. *Zhurn. Organ. Khimii* 2007, *43*, 1802–1806 [*Russ. J. Org. Chem.* 2007, *43* (Engl. Transl.)].

[155] Thanedar, S.; Farona, M. F. *J. Organomet. Chem.* 1982, *235*, 65–68.

[156] D'yakonov, V. A.; Galimova, L. F.; Ibragimov, A. G.; Dzhemilev, U. M. *Zhurn. Organ. Khimii* 2008, *44*, 1308–1312 [*Russ. J. Org. Chem.* 2008, *44*, 1291 (Engl. Transl.)].

[157] Rosenthal, U.; Oehme, G.; Burlakov, V. V.; Petrovsky, P. V.; Shur, V. B.; Vol'pin, M. E. *J. Organomet. Chem.* 1990, *391*, 119–122.

[158] U. Rosenthal, H. Görls, V. V. Burlakov, V. B. Shur, and M. E. Vol'pin, *J. Organomet. Chem.* 1992, *426*, C53–C57.

[159] Burlakov, V. V.; Polyakov, A. V.; Yanovsky, A. I.; Struchkov, Yu. T.; Shur, V. B.; Vol'pin, M. E.; Rosenthal, U.; Görls, H. *J. Organomet. Chem.* 1994, *476*, 197–206.

[160] Dzhemilev, U. M. et al. RF Patent 2162851; *Byull. Izobret.* [*Bulletin of Inventions*], 2001, No.4 (in Russian).

[161] Dzhemilev, U. M.; Ibragimov, A. G.; Yakovleva, L. G.; Khafizova, L. O.; Borisova, A. L.; Yakupova, L. R. *Izv. Akad. Nauk, Ser. Khim.* 2003, 1490–1499 [*Russ. Chem. Bull., Int. Ed.* 2003, *52*, 1573 (Engl. Transl.).

[162] Dzhemilev, U. M.; Sultanov, R. M.; Gaimaldinov, R. G.; Muslukhov, R. R. In Materialy Vsesoyuznoi Konferensii "*Primenenie metallokompleksnogo kataliza v organicheskom sinteze*", Tez. Dokl., Ufa, 1989 [Proceedings of the All-Union Conference "Application of Metal Complex Catalysis to Organic Synthesis" (Abstracts of Reports), Ufa, 1989] 40.

[163] Dzhemilev, U. M.; Sultanov, R. M.; Gaimaldinov, R. G.; Tolstikov, G. A. *Izv. Akad. Nauk SSSR, Ser. Khim.* 1991, 1388–1393 [*Bull. Acad. Nauk USSR, Div. Chem. Sci.* 1991, *40*, 1229 (Engl. Transl.)].

[164] Dzhemilev, U. M.; Sultanov, R. M.; Gaimaldinov, R. G.; Muslukhov, R. R.; Lomakina, S. I.; Tolstikov, G. A. *Izv. Akad. Nauk SSSR, Ser. Khim.* 1992, 980−999 [*Bull. Acad. Nauk USSR, Div. Chem. Sci.* 1992, *41*, 770 (Engl. Transl.)].

[165] Dzhemilev, U. M.; Sultanov, R. M.; Gaimaldinov, R. G. *J. Organomet. Chem.* 1995, *491*, 1−10.

[166] Dzhemilev, U. M.; Sultanov, R. M.; Gaimaldinov, R. G. *Izv. Akad. Nauk, Ser. Khim.* 1993, 165−169 [*Russ. Chem. Bull.* 1993, *42*, 149 (Engl. Transl.)].

[167] Takahashi, T.; Seki, T.; Nitto, Y.; Saburi, M.; Rousset, C. J.; Negishi, E. *J. Am. Chem. Soc.* 1991, *113*, 6266−6268.

[168] Hoveyda, A. H.; Xu, Z. *J. Am. Chem. Soc.* 1991, *113*, 5079−5080.

[169] Knight, K. S.; Waymouth, R. M. *J. Am. Chem. Soc.* 1991, *113*, 6268−6270.

[170] Dzhemilev, U. M.; D'yakonov, V. A.; Khafizova, L. O.; Ibragimov, A. G. *Zhurn. Organ. Khimii* 2005, *41*, 363−368 [Russ. J. Org. Chem. 2005, *41*, 352 (Engl. Transl.)].

[171] Lai, Y.-H. *Synthesis* 1981, 585−589.

[172] Grubbs R. H.; Miyashita, A. *Chem.Commun.* 1977, *23*, 864−866.

[173] Wisehmeyer, U.; Knight, K. S.; Waymouth, R. M. *Tetrahedron Lett.* 1992, *33*, 7735−7738.

[174] Knight, K. S.; Wang, D.; Waymouth, R. M. *J. Am. Chem. Soc.* 1994, *116*, 1845−1854.

[175] Waymouth, R. M.; Knight, K. S. *Chemtech* 1995, 15−20.

[176] Yamaura, Y.; Mori, M. *Tetrahedron Lett.* 1991, *40*, 3221−3224.

[177] Yamaura, Y.; Hyakutake, M.; Mori, M. *J. Am. Chem. Soc.* 1997, *119*, 7615−7616.

[178] Martin, S.; Brintzinger, H.-H. *Inorg. Chim. Acta* 1998, *280*, 189−192.

[179] Ojima, I.; Tzamarioudaki, M.; Li, Z.; Donovan, R. J. *Chem. Rev.* 1996, *96*, 635−662.

[180] Uesaka, N.; Saitoh, F.; Mori, M.; Shibasaki, M.; Okamura, K.; Date, T. *J. Org. Chem.* 1994, *59*, 5633−5642.

[181] Uesaka, N.; Mori, M.; Okamura, K.; Date, T. *J. Org. Chem.* 1994, *59*, 4542−4547.

[182] Mori, M.; Kakaki, T.; Sato, M.; Sato, Y. *Tetrahedron Lett.* 2003, *44*, 3797−3800.

[183] E. Negishi, In Comprehensive Organic Synthesis. Vol. 5. (Ed. by B.M. Trost). Pergamon. Oxford 1991, 1163.

[184] Rajanbabu, T. V.; Nugent, W. A.; Taber, D. F.; Fagan, P. J. *J. Am. Chem. Soc.* 1988, *110*, 7128−7135.

[185] Wender, P. A.; McDonald, F. E. *J. Am. Chem. Soc.* 1990, *112*, 4956–4958.

[186] Agnel, G.; Negishi, E. *J. Am. Chem. Soc.* 1991, *113*, 7424–7426.

[187] Agnel, G.; Owczarczyk, Z.; Negishi, E. *Tetrahedron Lett.* 1992, *33*, 1543–1546.

[188] Mori, M.; Uesaka, N.; Skibasaki, M. *J. Org. Chem.* 1992, *57*, 3519–3521.

[189] Lund, E. C.; Livinghouse, T. *J. Org. Chem.* 1989, *54*, 4487–4488.

[190] Dzhemilev, U. M.; D'yakonov, V. A.; Khafizova, L. O.; Ibragimov, A. G. *Tetrahedron* 2004, *60*, 1287–1291.

[191] D'yakonov, V. A.; Zinnurova, R. A.; Ibragimov, A. G.; Dzhemilev, U. M. *Zhurn. Organ. Khimii* 2007, *43*, 962–966 [*Russ. J. Org. Chem.* 2007, *43*, 956 (Engl. Transl.)].

[192] D'yakonov, V. A.; Makarov, A. A.; Ibragimov, A. G.; Dzhemilev, U. M. *Zhurn. Organ. Khimii* 2008, *44*, 207–211 [*Russ. J. Org. Chem.* 2008, *44*, 197 (Engl. Transl.)].

[193] D'yakonov, V. A.; Makarov, A. A.; Ibragimov, A. G.; Khalilov, L. M.; Dzhemilev, U. M. *Tetrahedron* 2008, *64*, 10188–10194.

[194] Dzhemilev, U. M.; Ibragimov, A. G. Uspekhi Khimii 2005, *74*, 886–904 [*Russ. Chem. Rev.* 2005, *74*, 807 (Engl. Transl.)].

[195] Dzhemilev, U. M.; Ibragimov, A. G.; D'yakonov, V. A.; Zinnurova, R. A. *Zhurn. Organ. Khimii* 2007, *43*, 184–188 [*Russ. J. Org. Chem.* 2007, *43*, 176 (Engl. Transl.)].

[196] D'yakonov, V. A.; Makarov, A. A.; Dzhemilev, U. M. *Zhurn. Organ. Khimii* 2009, *45*, 1612–1617 [*Russ. J. Org. Chem.* 2009, *45* (Engl. Transl.)].

[197] Rousset, C. J.; Negishi, E.; Suzuki, N.; Takahashi, T. *Tetrahedron Lett.* 1992, *33*, 1965–1968.

[198] Negishi, E.; Takahashi, T. *Acc. Chem. Res.* 1994, *27*, 124–130.

[199] Negishi, E.; Takahashi, T. *Bull. Chem. Soc. Jap.* 1998, *71*, 755–769.

[200] Lewis, D. P.; Whitby, R. J.; Jones, R. V. H. *Tetrahedron* 1995, *51*, 4541–4550.

[201] Rousset, C. J.; Swanson, F. L.; Lamaty, F.; Negishi, E. *Tetrahedron Lett.* 1989, *30*, 5105–5108.

[202] Negishi, E.; Maye, J.P.; Choueiry, D. *Tetrahedron* 1995, *51*, 4447–4462.

[203] Negishi, E.; Rousset, C.J.; Choueiry, D.; Maye, J. P.; Suzuki, N.; Takahashi, T. *Inorg. Chim. Acta* 1998, *280*, 8–20.

[204] Walker, F. W.; Ashby, E. C. *J. Am. Chem. Soc.* 1969, *91*, 3845–3850.

[205] Khalilov, L. M.; Parfenova, L. V.; Rusakov, S. V.; Ibragimov, A. G.; Dzhemilev, U. M. *Izv. Akad. Nauk, Ser. Khim.* 2000, 2086–2093 [*Russ. Chem. Bull.* 2000, *49*, 2051 (Engl. Transl.)].

[206] Rusakov, S. V.; Khalilov, L. M.; Parfenova, L. V.; Ibragimov, A. G.; Ponomarev, O. A.; Dzhemilev, U. M. *Izv. Akad. Nauk, Ser. Khim.* 2001, 2229−2238 [*Russ. Chem. Bull., Int. Ed.* 2001, *50*, 2336 (Engl. Transl.)].

[207] Balaev, A. V.; Parfenova, L. V.; Gubaidullin, I. M.; Rusakov, S. V.; Spivak, S. I.; Khalilov, L. M.; Dzhemilev, U. M. *Dokl. Akad. Nauk* 2001, *381*, 364−367 [*Dokl. Chem.* 2001, *381*, 279 (Engl. Transl.)].

[208] Parfenova, L. V.; Gabdrakhmanov, V. Z.; Khalilov, L. M.; Dzhemilev, U. M. *J. Organomet. Chem.* 2009, *694*, 3725−3731.

[209] Dzhemilev, U. M.; Ibragimov, A. G.; D'yakonov, V. A.; Pudas, M.; Bergmann, U.; Khafizova, L. O.; Tyumkina, T. V. *Zhurn. Organ. Khimii* 2007, *43*, 686−689 [*Russ. J. Org. Chem.* 2007, *43*, 681 (Engl. Transl.)].

[210] Dzhemilev, U. M.; D'yakonov, V. A. et al. RF Patent 2269505; Byull. Izobret. [Bulletin of Inventions], 2006, No. 4 (in Russian).

[211] Khafizova, L. O.; Gilyazev, R. R.; Tyumkina, T. V.; Ibragimov, A. G.; Dzhemilev, U. M. *Zhurn. Organ. Khimii* 2007, *43*, 967−671 [*Russ. J. Org. Chem.* 2007, *43*, 961 (Engl. Transl.)].

[212] D'yakonov, V. A.; Makarov, A. A.; Dzhemilev, U. M. *Tetrahedron* 2009, (in press).

[213] D'yakonov, V. A. et al. (unpublished results).

Usein M. Dzhemilev received his Candidate (Ph.D.) and Doctor of Science (Chemistry) degrees from the Institute of Chemistry, Bashkirian Branch of Academy of Sciences of the USSR (nowadays the Institute of Organic Chemistry of Ufa Scientific Centre of RAS). In 1982, he became a professor, and since 1980 he has held the position of the Deputy Director of the same Institute. Since 1992 he has become Director of the Institute of Petrochemistry and Catalysis of Bashkortostan Republic Academy of Sciences, which since 2004 has associated with the Russian Academy of Sciences. As a visiting scientist he reported in Poland, Czechoslovakia, Bulgaria, Israel, China and USA.

In 1990 he was elected Corresponding Member of the Russian Academy of Sciences. He is a Laureate of State Prizes in the area of science and technology of the USSR (1990) and the Russian Federation (2004). His research interests include metal complex catalysis in the organic and organometallic synthesis, chemistry and stereochemistry of the strained and cage compounds, organic chemistry of nontrasition metals (Mg, Al, Zn, Ga, In) as well as chemistry of small, low-stable and highly strained molecules.

He is the author and co-author of three books, about 1500 scientific articles and reviews. He also has more than 600 patents issued.

INDEX

A

acceptor, 8, 26, 42, 43, 62, 83
acetate, 35
acetylene, ix, 3, 46, 47, 49, 50, 53, 54, 64, 67, 77, 84
achievement, 49, 52
activation, 8
adducts, 29
agents, 86
aid, 12, 13, 17, 24, 28, 33, 51, 52, 54, 60
alcohols, 19, 20, 37, 38, 43
aldehydes, 37
alkanes, 5, 21
alkenes, 17, 18
alkylation, 20, 47
aluminum, 10, 12, 13, 24, 26, 30, 31, 34, 69
amines, 19, 21
application, ix, 1, 10, 18, 21, 23, 24, 32, 44, 87
atoms, ix, 1, 2, 8, 10, 13, 19, 32, 76, 83
availability, 2, 23

B

beetles, 38
benzene, 36, 57

binuclear, 72
biologically active compounds, 23
birth, 1
Blattella germanica, 38
bonds, 17, 18, 19, 23, 33, 34, 43, 51, 67, 77, 82, 85
butadiene, 42
butane, 11, 39

C

carbon, 2, 23, 32, 83
carbon atoms, 32, 83
carboxylates, 37
catalyst, v, 5, 6, 11, 19, 24, 25, 27, 28, 29, 31, 33, 34, 35, 37, 42, 44, 45, 47, 48, 49, 50, 51, 52, 55, 57, 58, 59, 61, 62, 64, 65, 69, 70, 72, 76, 81, 82, 83, 85, 104
chiral, 19, 24, 58
chloride, 5, 34, 45, 65, 77, 84
cis, 58, 73, 82
classes, ix, 12, 13, 14, 46, 67, 87
cleavage, 33, 71, 72, 84, 85
cockroach, 38
communication, 47, 69
composition, 5, 85
compounds, v, ix, x, 1, 2, 3, 5, 8, 12, 13, 14, 17, 19, 23, 28, 32, 38, 39, 41, 44, 46, 54,

56, 58, 60, 62, 67, 69, 74, 75, 79, 81, 82, 84, 87, 104
concentration, 5
configuration, 14, 28, 32, 57
congress, v
conjugated dienes, 58
construction, 10, 11, 45, 81, 86
conversion, 60
coupling, 34, 84, 85
criticism, 11
cyclohexane, 44

D

data collection, 87
decane, 55
decoding, 75
deficiency, 30
derivatives, 7, 8, 10, 13, 59, 69, 81
dichloroethane, 33, 49
dienes, ix, 1, 3, 5, 24, 34, 35, 43, 45, 46, 50, 55, 58, 59, 62, 67, 72, 87, 88
dimeric, 74
dimerization, 5, 6, 11, 76
diversity, 23
donor, 8, 26, 32
double bonds, 17, 18, 43, 77, 82, 85
duration, 53

E

elaboration, 12, 88
electron pairs, 26
environment, 26, 76
equilibrium, 1
ethane, 76, 77
ethers, 19, 20, 71
ethylene, 10, 20, 33, 49, 70
exaggeration, 2, 88
experimental condition, 61
exploitation, 86

F

family, 14
fire, 32
flow, 67
fullerene, 29, 37, 42

G

generation, 6, 8, 28, 31, 34, 42, 57, 60, 70, 75
Grignard reagents, 3, 17, 21, 58, 62, 74, 82, 84
groups, 7, 57, 72

H

halogen, 2, 87
heating, 62
heptane, 18, 41
heterocycles, x, 3, 14, 34
hexane, 44, 48
hydride, 34, 76
hydro, 39, 62, 64, 85
hydrocarbons, 39, 62, 82, 85
hydrogen, 9, 32, 34, 77
hydrogen atoms, 9
hydrolysis, 18, 32, 38, 82

I

identification, 75
implementation, 87
in situ, 3, 23, 33, 34, 36, 37, 38, 40, 42, 65, 87
industry, 86
inert, 13
initial reagents, 31, 56, 58, 71, 73, 87
injury, v
insertion, 72, 76
insight, 23

interaction, 1, 8, 9, 11, 18, 27, 29, 30, 34, 36, 37, 42, 44, 48, 56, 57, 58, 61, 71, 74, 76, 77, 83
intermolecular, 45, 51, 52, 62, 64, 82, 83
ions, 42, 44, 62, 83
isolation, 84, 86, 87
isomers, 32

K

ketones, 37, 84
kinetic model, 76

L

ligand, 6, 8, 9, 76
limitation, 2, 30
linear, 5

M

macromolecules, 86
magnesium, 1, 10, 49, 61, 69, 74
magnetic, v
media, 59
medicine, 86
metals, v, ix, 1, 2, 6, 9, 13, 24, 86, 104
methylene, 5, 40, 43, 44
moieties, 29
molar ratio, 35
mole, 11, 27, 30, 35, 39, 43, 44, 45, 48, 53, 55, 59, 61, 62, 64, 72, 82, 83, 84
molecules, 82, 104
monomeric, 74
monomers, 14, 23, 86

N

natural, 21, 23
NMR, 32, 47, 60, 75, 85
noble metals, 86

nonane, 40
norbornene, 28, 57
normal conditions, 2

O

OAC, 27, 28, 30, 32, 33, 37, 45, 46, 52, 53, 76, 83
octane, 40, 41, 74
organ, v, ix, 2, 8, 14, 17, 23, 46, 67, 69, 82, 87, 88, 104
organic, v, ix, 2, 12, 13, 14, 46, 60, 67, 69, 88, 104
organic compounds, ix, 12, 13
organoaluminum compounds, x, 79, 84
organomagnesium, x, 2, 18, 19, 56, 57, 58, 67, 73, 79, 84, 86, 87
organomagnesium compounds, 19, 67, 87
organometallic, i, ii, iii, v, ix, 2, 8, 14, 17, 23, 46, 67, 69, 82, 87, 88, 89, 95, 97, 104
organometallic chemistry, 46, 88
oxidation, 19, 24, 25
oxidation products, 19
oxidative, 8, 84, 85

P

patents, 104
pathways, 87
pentane, 40
physicochemical properties, 12
polyketones, 85
polymeric macromolecules, 86
polymerization, 44, 69
polymerization mechanism, 69
press, 98, 102
property, v
propylene, 35
purification, 86

Q

quantum, 76

R

range, 18, 23
reactant, 57
reaction temperature, 56, 73
reactivity, 2, 33, 34, 46, 67
reagents, 1, 2, 3, 11, 14, 17, 20, 21, 23, 29, 31,
    32, 56, 57, 58, 62, 65, 71, 73, 74, 76, 81,
    82, 84, 85, 86, 87
recognition, 2
regeneration, 70, 72, 76
regioselectivity, 20, 44
retention, 48, 57
room temperature, 13, 35
Russian Academy of Sciences (RAS), 103, 104

S

safety, 32
salts, 24, 37, 60
selectivity, 11, 20, 24, 28, 31, 35, 36, 58, 65,
    73
selenium, 36
separation, 86
series, 66
silane, 27
solvent, 20, 25, 32, 43, 44, 56, 58, 64, 65, 71,
    72, 73, 74, 76, 84
spectroscopic methods, 85
spectroscopy, 47, 75
spectrum, 23
stability, 2, 12
stages, 72, 76, 77
styrene, 55, 57
substances, 1, 21
sulfur, 62, 65, 66

synthesis, v, ix, 1, 2, 3, 11, 12, 21, 23, 24, 30,
    34, 37, 38, 39, 41, 42, 46, 47, 49, 59, 64,
    67, 81, 84, 85, 86, 87, 88, 104

T

technology, 104
temperature, 13, 20, 30, 35, 49, 56, 59, 61,
    65, 71, 73
tetrahydrofuran, 20, 30, 33, 74
tetrahydrofurane, 49
thermodynamic stability, 12
time, 7, 18, 39, 47, 57, 59, 61
titanium, 13, 17, 21, 52, 93
toluene, 29, 42, 50
trans, 10, 30, 31, 32, 33, 35, 36, 37, 50, 58,
    73, 83
transfer, v, 8, 32, 34, 74, 76, 77
transformations, 1, 2, 3, 7, 8, 10, 11, 23, 43,
    47, 48, 58, 66, 70, 71, 72, 74, 75, 76, 87
transition, ix, 1, 2, 6, 8, 9, 12, 13, 17, 24, 86
transition metal, ix, 1, 2, 6, 8, 13, 17, 24, 86

V

valence, 35, 52
versatility, 23

X

xylene, 66

Y

yield, 5, 18, 19, 25, 27, 30, 31, 32, 34, 35, 40,
    42, 43, 44, 46, 50, 52, 53, 55, 56, 57, 61,
    62, 63, 64, 65, 73, 74, 82, 83

Z

zirconium, 13, 69, 71, 74